30 MORE MATH MYSTERIES
Kids Can't Resist!

Quick & Clever Mysteries That Boost Problem-Solving Skills

**Martin Lee With
Marcia Miller**

■SCHOLASTIC

New York • Toronto • London • Auckland • Sydney
Mexico City • New Delhi • Hong Kong • Buenos Aires

Dedication

To Joansie, with love & tail wags

Editor: Maria L. Chang
Cover design by Michelle H. Kim
Cover illustration by John Lund
Interior design by Maria Lilja
Interior illustrations by Matt Ward
Photos © The Noun Project: 8 key and throughout (yanti),
8 magnifying glass and throughout (Danilo Gusmão Silveira),
9 magnifying glass and throughout (Russell Shaw).

Scholastic Inc., 557 Broadway, New York, NY 10012
ISBN: 978-1-338-25730-4
Copyright © 2019 by Martin Lee
All rights reserved.
Printed in the U.S.A.
First printing, January 2019.

1 2 3 4 5 6 7 8 9 10   40   25 24 23 22 21 20 19

Contents

Math Mysteries

Introduction

Don't you love the thrill of solving a knotty mystery? Most kids do. They enjoy challenges and usually respond to them with enthusiasm. They typically experience a deep feeling of accomplishment upon grasping a complex idea or solving a tricky problem.

Solving mysteries is all about identifying and unraveling problems. According to the Common Core State Standards for Mathematics (CCSS) and most state standards, problem solving is essential to inquiry and to application, and should be interwoven throughout the math curriculum at all grade levels. Math educators agree that students benefit from frequent experiences in working independently and collaboratively to solve problems. They believe that well-chosen problems can be uniquely valuable in developing or deepening students' understanding of powerful mathematical ideas.

Good problems are those that students can access on different levels and that challenge them in a variety of ways. Good problems get kids to reason mathematically. They require students to apply number sense, logic, and intuitive thinking skills; to identify and explore patterns; to make and test reasonable guesses; to show flexibility in thinking; to adjust their assumptions; to work backwards; to use manipulatives; to make sketches or tables; or even to act out situations. The best problems build on and extend students' mathematical language, skills, and understanding as they lead students to develop and sharpen reasoning and communication skills. They afford students the chance to gauge their own strengths and weaknesses as problem solvers.

As students tackle the mysteries in this book, they can turn into math detectives. As such, they'll need to bring all their skills to an investigation. First, they will need to focus their attention on each mystery, evaluating what they know and determining what additional information they might need. Then they must determine a solution strategy that fits the problem. Next, students will apply that strategy to obtain a solution, and then verify whether their solution makes sense. Finally, they will reflect to decide what conclusions they can draw from their findings and how best to communicate their findings accurately, using precise vocabulary. As students work their way through these steps, they learn to be observant, creative, and flexible in their approach to new situations.

Providing students with motivating mysteries to unravel does more than give them the opportunity to feel good—it will help them experience the power, utility, and elegance of mathematics. It will give them the chance to apply and adapt a variety of appropriate strategies to solve real-world problems—and this is exactly what today's higher standards require.

The eight Standards for Mathematical Practice identified in the CCSS resource materials summarize goals you can help instill in your students. The mysteries in this book offer them the opportunity to strive toward these goals as they wrestle with solving real-world and non-routine math problems.

These overarching goals are not particular skills but rather broad practices. They belong in any math class in support of all math curricula. You may wish to post these practices (reword, as needed, in age-appropriate language) to inspire students to become successful, confident problem solvers.

- Make sense of problems and persevere in solving them.

- Reason abstractly and quantitatively.

- Construct viable arguments and critique the reasoning of others.

- Model with mathematics.

- Use appropriate tools strategically.

- Attend to precision.

- Look for and make use of structure.

- Look for and express regularity in repeated reasoning.

I hope you'll find that the problems in *30 More Math Mysteries Kids Can't Resist!* engage your students and enrich your math curriculum. Happy detecting!

Correlations to Mathematics Standards

Mystery	Operations & Algebraic Thinking	Number & Operations in Base Ten	Number & Operations— Fractions	Measurement & Data	Geometry
Time and Temperature	✔	✔		✔	
Dollars and Sense	✔	✔		✔	
Huh?					
Ball Bandit				✔	✔
Too Little Information	✔	✔		✔	✔
Too Much Information	✔	✔	✔	✔	
Doggy Dilemma	✔			✔	
Puzzling Payment Plan	✔	✔		✔	
Code Crackers	✔			✔	✔
Musical Chairs	✔	✔			
Suki's Salamander	✔	✔		✔	✔
Fountain of Wishes	✔	✔		✔	
Digit Goes to the Movies	✔			✔	
Safer Safe	✔	✔			
Carpet Confusion	✔	✔		✔	✔
Taco Take-Away	✔	✔	✔		
Fair and Squares	✔	✔		✔	✔
It's a Wrap!	✔	✔		✔	✔
Forgettable Flavors	✔	✔			
Turkey Talk	✔	✔	✔		
And the Final Score Was…	✔	✔	✔		
Cat Food Case	✔	✔	✔		
Hiking Head-Scratcher	✔	✔	✔	✔	
Party Animals	✔	✔			
Relative Ages	✔	✔			
Finders, Keepers	✔	✔	✔	✔	
Tech Time-Crunch	✔	✔	✔	✔	
Fish Scales	✔	✔	✔	✔	
Meeting Marvin	✔	✔		✔	
The Broom King	✔	✔		✔	✔

Using This Book

- This book opens with an introduction to the detective team, followed by 30 reproducible math mysteries. Not every mystery may be appropriate for every class or student. Use the math standards correlations chart (p. 7) to help you decide whether a mystery is right for your class. Pick and choose as you see fit. Feel free to adapt or adjust problems to suit the ability levels and interests of your students.

- Teaching notes for each mystery begin on page 10. These notes provide full answers, useful pointers, and math background information, as appropriate. Use this section to learn more about the skills and concepts involved in each solution.

- Good detectives need to be not only good observers and listeners but also good readers and summarizers. These mysteries combine elements of mathematics and language arts, as well as other curriculum connections. When students discuss their approaches and solutions to a mystery, encourage them to tell not only what clue(s) they found but also where and how they detected it (them).

- Students may not regard solving math mysteries as reading assignments, but they are! Guide students to read each story thoroughly and carefully, actively engage with the text, and not just seek numbers to pluck out and manipulate. Encourage them to employ the close-reading strategy of text marking and to recognize and highlight key terms and important details. Invite them to make notes on the margins, record comments or questions on sticky notes, or use the reproducible Investigator's Log (p. 25).

- Another successful comprehension strategy is to have students first do a quick read of each story for its gist. Then have them reread the mystery more slowly and analytically, marking the text as they go. Also, you might draw attention to the titles, as some hint at the theme of the mystery, while others are humorous, alliterative, or twists on familiar phrases and idioms.

- Some problems may present unfamiliar situations. As needed, allow time to discuss such circumstances with students. Also, talk about tasks to complete and clarify any words, expressions, or idioms that may stump them.

Call attention to the recurring icons that support each mystery

The key icon alerts students to the specific problem at hand or to a key math concept or strategy upon which the solution rests.

The magnifying glass icon clues in readers that information in that paragraph or portion of the text requires especially close and careful reading. If you wish to provide a greater challenge, you might mask the icons before you reproduce the mystery.

- Another way to help students access the text is to have them bring their prior knowledge and experiences to the discussion—as true detectives do. You can also invite them to retell the tales in their own words, or have small groups act out a mystery for the class.

- Each mystery requires students to combine arithmetic skills, solution strategies, and logical/mathematical reasoning. Few can be solved quickly or directly using one or more arithmetic operations alone. Help students appreciate that there's not necessarily one way to solve a given mystery; in some cases, there's not only a single right answer. Inspire them to be flexible, to try different problem-solving strategies, to "think outside the box," and to be patient, but tenacious.

- Try tackling these mysteries yourself to engage students in the process. They will benefit from observing as you behave like a math detective. Model how to approach and "unpack" a non-routine problem, including ways to unearth the clues built into every story.

- Vary the way you present the mysteries. You might display them on a bulletin board, distribute copies of them, post them on a whiteboard or class website, or read some aloud.

- Many of the mysteries lend themselves to collaborative work. Determine the best grouping to suit the learning styles of your students. Encourage them to share solution strategies and real-life mathematical reasoning. Guide students to acknowledge and respect others' diverse ideas, presentations, or solution pathways.

- An Investigator's Log (p. 25) follows the teaching notes. Reproduce it for students so they can record information they gather in the process of untangling a mystery. You may want to revise the log to include other questions you think are important for students to answer for any given mystery, or customize it to fit a particular one more closely. Consider having students compile their completed Investigator's Logs in an investigator's portfolio.

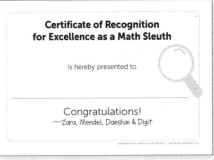

- Solving these mysteries may stimulate students' taste for sleuthing. Consult your school librarian to identify books containing math mysteries, such as the Encyclopedia Brown or the 39 Clues series. Invite students to share math mysteries they know or have read.

- Fill out a Certificate of Recognition (p. 80) to acknowledge the efforts of each classroom sleuth. If possible, plan a sleuthing party in which you set up a scavenger hunt or other mystery task to celebrate students' progress as math detectives.

Teaching Notes

Meet the Detectives (p. 26)

- Although there is no mystery to discuss or solve for this section, it's worth taking time to talk with students about their impressions of detectives Zara and Mendel, as well as of Daeshim and Digit. Ask students: *What does the tone of this introduction tell you about what you might expect in the upcoming mysteries?*

- Invite students to share their ideas about the characteristics of good detectives or problem solvers. Have them suggest particular skills and traits that successful sleuths often share. If no one mentions it, point out that good listening skills and good questioning skills are essential to successful sleuthing. Encourage students to notice as they read how Zara and Mendel exhibit these attributes.

- After students work through the mysteries, invite them to write to Zara or Mendel to offer advice, congratulations, alternate solutions or strategies, or constructive criticisms. Digit loves getting mail, too!

- Clarify that the preteen detectives do not capture criminals nor pursue perpetrators of serious crimes—those responsibilities belong to law-enforcement personnel. Tell them that the "crimes" the detectives face in this book have a decidedly lighter tone. Remind them that no dogs or other animals have been harmed by the detectives' actions.

Time and Temperature (p. 28)

Math Skills/Concepts: measurement (intervals of time, temperature), represent and interpret data, patterns and relationships

Teaching Tips

- Guide students to understand that the text in italics represents the text messages that went back and forth between James and Mendel.

- Confirm that students understand Granny's strict swimming rule.

- Ask: *Why do you think Mendel wanted to know when James checked the temperature and what it was each time?* (To gather more information and perhaps look for patterns in the data)

- Guide students to look for a pattern in the temperature changes.

- Help students recognize that the table is a graphic representation of the time and temperature data Mendel collected from James. Invite students to extend the pattern until the solution reveals itself.

- Explain to students that although temperatures do not conveniently rise or fall according to a clear pattern in real life, they often change in predictable ways. This lesson is meant to demonstrate the idea that recognizing a pattern is a very useful problem-solving strategy.

- Reinforce or extend by presenting other problems in which the solutions depend on recognizing and continuing number patterns.

Suggested Solution: If the pattern continues, the brothers can swim at 10:30 a.m., when the temperature will reach 80°F. Mendel reached this solution by detecting and applying a pattern of temperature rise per half-hour as +6°, +5°, +4°, and so on.

Dollars and Sense (p. 30)

Math Skills/Concepts: money computation, rounding money, multistep problem, number sense

Teaching Tips

- 🔍 This section provides the rules for choosing party favors. Suggest that students mark the text to highlight the word *different* and the number of party guests.
- 🗝 Focus students on the tricky aspects of the problem: Buster can spend no more than $5 but wants to come as close to $5 as possible. Again, encourage students to mark these details on the page.
- Discuss these questions with students: *How can Buster decide which items to pick? Are there ones he can rule out right away? How do you know?*
- Some students may be able to solve this problem using estimation—rounding to the nearest dime. Invite a volunteer to demonstrate this strategy to the class.
- Extend by providing similar types of shopping-decision problems for students to solve.

Suggested Solution: The best one is bolo tie + saddle/hat pin + cowboy boot keychain + cactus pen = $4.96. A second set of four gifts (steer figurine + saddle/hat pin + cowboy boot keychain + cactus pen = $4.76) is also under $5 but not as close to it.

Huh? (p. 31)

Math Skills/Concepts: logical reasoning, creative thinking

Teaching Tips

- 🔍 Have students read this paragraph twice to ensure they understand Daeshim's claim.
- 🗝 Discuss the meaning of the expression "powers of observation." Challenge students to imagine themselves as Daeshim in a park and suggest what might surround him as he tosses the ball into the air.
- Unlike most mysteries in this book, this one has no set answer. Invite as many original solutions as students can generate. Encourage them to "think outside the box."
- This problem lends itself to group thinking. You can solve it together as a whole class or put students into small groups to brainstorm and share ideas.

Suggested Solution: Possible ideas—the ball landed in tree or bush; it fell into a lake or fountain; another person (or dog) caught it; it landed atop a park building/table/bench; it rolled into a pipe or other hole; a bird in flight snatched it.

Ball Bandit (p. 32)

Math Skills/Concepts: geometry (parallel/perpendicular, right angles), mapping, compass directions

Teaching Tips

- ☞ Guide students to understand that a street map is a two-dimensional representation of a place as seen from above—as if from a bird's-eye view.
- 🔍 **1)** Encourage students to mark the text to highlight each direction and distance.
  - **2)** Have students sketch Pokey's route right on the street map, and then check that their sketch matches the given directions. Note: Some students might mistake the term *right-angle turn* for *turn right*. Guide them to understand the difference between the two phrases.
- ● Help students review and apply specific math terminology as they describe the street map: *parallel, perpendicular, intersection, right angle, midpoint* (mid-block). Model how to include these terms.
- ● Invite students to think of other locations on the map where Pokey might hide a ball. Then have them challenge classmates to follow their precise directions to find the ball.
- ● If appropriate, extend by adjusting the map: insert and name one or two diagonal streets.

Suggested Solution: Pokey buried Daniel's ball on Cheetah Court about halfway between Lilac Street and Marigold Street. The solution is based on understanding right angles, directions left and right, and north, south, east, and west.

Too Little Information (p. 34)

Math Skills/Concepts: measurement, money computation, logical reasoning

Teaching Tips

- ☞ Guide students to recognize that Kenta is part of the problem because he tends to omit some information.
- 🔍 This paragraph reveals some important information about Kenta's situation, including data about the size of Mr. Weedler's lawn.
- ● As needed, guide students to recognize how the data about how much Mr. Weedler pays Kenta and the size of his lawn relates to Ms. Greengrass's question.
- ● Challenge students to explain why Kenta's estimate of 45 minutes until he finishes is extra information. Point out that readers don't know when Kenta started the job. It's important to consider the size of the two lawns, not the time it takes him to finish mowing one.

Suggested Solution: The missing information is the size of Ms. Greengrass's lawn so the detectives can compare it to the size of Mr. Weedler's lawn and help Kenta set a fair price for the job.

Too Much Information (p. 35)

Math Skills/Concepts: multistep problem, multiplication, time, fractions (½)

Teaching Tips

⚷ Discuss with students why extra information can make a problem more difficult to solve than it really is. Clarify that problem solvers need to differentiate between key facts and unnecessary ones.

🔍 This paragraph identifies the specific problem Ruth must answer. If necessary, remind students that 30 seconds = ½ minute.

● Discuss why Mendel asks Ruth, "How far along are you in your work?" Explain that this question helps focus Ruth's thinking on how much longer she'll need to work to complete the original task.

● Extend by challenging students to determine how many nails Ruth needed to shutter all the windows in all the cabins. *(10 cabins = 20 windows; each window needs 6 boards × 4 nails per board = 20 × 6 × 4 = 480 nails for all 20 windows)*

Suggested Solution: Ruth should tell her dad that she'll be done in 12 minutes, because one window needs 6 boards and each board needs 4 nails. 6 × 4 = 24 nails, and 24 nails × ½ minute per nail = 12 minutes. The number of cabins and the number of windows and doors each cabin has are extraneous information.

Doggy Dilemma (p. 36)

Math Skills/Concepts: patterns, common multiples, creative thinking

Teaching Tips

🔍 **1)** Have students restate in their own words the nature of Luis's problem.

2) Careful reading of this paragraph is essential. Students should recognize that each dog's schedule follows a pattern and that visualizing the relationship between the patterns will help Luis solve his problem. Have them highlight the italicized terms as descriptors of the patterns.

⚷ The key problem-solving strategy here is to use a calendar or chart to determine which days both dogs would be at the dog park—something Luis wants to avoid.

● Discuss these questions: *On which day were both dogs at the park at the same time?* (Monday) *Why did Zara ask Luis about a calendar? How could Luis use it to solve his problem?* (Luis can mark the days each dog goes to the park according to their schedule and see when the days coincide.)

● Challenge students to think of a way for Luis to bring Lola and his cousins to the dog park on the weekend and avoid Jasper. One way is to adjust Lola's schedule and bring her to the park on Saturday or on Sunday morning. Or, Luis might ask Jamie to switch Jasper's schedule a bit.

Suggested Solution: If both dog owners stick to their schedules, Luis can't safely bring Lola and his cousins to the dog park the following weekend. Both dogs were at the park on Monday. According to their patterns, Lola will be at the park next on Wednesday, Friday, and Sunday afternoons. Jasper's pattern takes them to the park next on Thursday and Sunday afternoons. Based on these schedules, Luis cannot bring his cousins to the dog park on Sunday afternoon.

Puzzling Payment Plan (p. 38)

Math Skills/Concepts: number sense, number patterns, whole-number operations

Teaching Tips

- 🔍 Careful reading of this paragraph is essential to understand the two payment plans Della must compare. Have students highlight this important data.
- 🔑 The most effective pre-algebraic strategy for solving this problem is to make and complete a table for each payment plan, and compare the results.
- Have students read the labels above each table to make sure they understand how the two differ. Invite them to explain why Mendel set up the tables as he did. Then, have students complete the tables on their own. Ask: *How should you use this data to answer Della's question?* (Find the sum of the pay for all eight walks under each plan.)
- Challenge students to identify data in the problem that is unnecessary to its solution (weights of dogs, length of time per walk).
- Extend by inviting a dog walker to visit your class and describe his or her experiences.

Suggested Solution: The doubling money plan will be much better for 8 walks; 8 walks at $20 each = $160; but the doubling plan would have a total of $1 + $2 + $4 + $8 + $16 + $32 + $64 + $128 = $255.

Code Crackers (p. 40)

Math Skills/Concepts: spatial reasoning, shape recognition, one-to-one correspondence

Teaching Tips

- Before reading this story, discuss with students what codes are, who might use codes, and why they can be so effective and fascinating. They might be aware of Morse code (letters formed by patterns of short and long sounds) or Braille (an alphabetic/numeric code to enable blind people to read). Students who can read music will understand that musical notation itself is a form of code.
- 🔑 As needed, guide students to recognize that each of the nine small sections of a 3-by-3 tic-tac-toe board has a distinct outline or shape. Ollie used these shapes as a code to spell out three words.
- 🔍 Without giving away the code, encourage students to look carefully at the tic-tac-toe board with letters in each section and the separate shapes that spell out a message. The trick is to match each shape in the coded message with its position within the tic-tac-toe board.
- Extend by inviting interested students to make up their own tic-tac-toe codes to spell their names, short phrases, or any word(s) that can be spelled with nine distinct letters.

Suggested Solution: Each of the nine distinct code shapes represents a different letter. By matching each tic-tac-toe shape with the letter it stands for, the code reads BEHIND THE SHED. That's where Ollie buried Digit's bone.

Musical Chairs (p. 42)

Math Skills/Concepts: multistep problem, whole-number operations

Teaching Tips

- ⚷ Here, the basic problem is presented: Kevin must set up 200 folding chairs according to a set of rules.
- 🔍 **1)** Close reading of this paragraph will enable students to learn the rules Ms. Basso has set. Encourage students to highlight the key data here.
  2) This paragraph specifies what to do with the leftover chairs.
- In the story, Ms. Basso sounds like a demanding person. Discuss with students why it might make sense to arrange rows with different numbers of seats in them. (The room may not be rectangular in shape; the production might require extra room for performers, recording equipment, etc.; staggered rows may create better sight lines; and so on.)
- This story includes hidden information some students may miss but need to include in the solution: the number of remaining rows *(4)* and the number of chairs left over *(14)*.
- Some students may be able to solve this problem more easily using manipulatives. Provide counters or snap cubes they can use to represent the chairs.

Suggested Solution: The first 3 rows have 10 chairs each *(30)*; the next 8 rows have 12 chairs each *(96)*, for 11 rows. The last 4 rows have 15 chairs each *(60)*. The total number of chairs in rows is 186, leaving 14 chairs to go along the side walls, 7 chairs per wall.

Suki's Salamander (p. 44)

Math Skills/Concepts: measurement (area, perimeter), geometry, factors

Teaching Tips

- 🔍 Have students carefully read this paragraph to help them visualize the general shape of the pen in relation to the size of the salamander.
- ⚷ This paragraph hints at the mathematical concept behind the solution—using factors to form rectangles of different shapes.
- As needed, review the meanings of the terms *perimeter* and *area*, and the formula for the area of a rectangle ($A = l \times w$). Using visual models, review the concept that rectangles of different dimensions can have the same area.
- Review the definition of *factor* (a number that can be divided evenly into a larger number). If helpful, work together to list all the factors of small numbers, such as 6, 10, and 16.

Suggested Solution: Sally's pen should be 3 inches wide and 28 inches long; $3 \times 28 = 84$; 3 is the smallest factor of 84 that is greater than 2½.

Fountain of Wishes (p. 45)

Math Skills/Concepts: multistep money problem, algebraic reasoning, number sense

Teaching Tips

- 🔑 Help students recognize that the task is more than determining the total value of the coins.
- 🔍 **1)** Have students closely read this sentence to determine the total value of the money collected from the fountain.
  - **2)** Guide students to highlight the key relationship they must use to divide up the money.
- ● The first part of this problem involves finding the grand total of the pennies, nickels, dimes, and quarters. Invite students to share different strategies for finding these values.
- ● To determine how to allocate the funds, some students may use their number sense to guess and test until they find the solution. Others may use algebraic reasoning: a number + twice that number = 24.

Suggested Solution: The money should be divided as follows: $8 for grow lamps and $16 for e-readers. To determine this, first find the total value of all the coins *($24)*, then figure out how to split $24 into two parts, the greater of which is twice the other.

Digit Goes to the Movies (p. 47)

Math Skills/Concepts: elapsed time, logical reasoning

Teaching Tips

- 🔑 Guide students to notice that the solution to this problem is based on what's best for Digit.
- 🔍 **1)** Have students carefully read the schedule to notice that the movies have different themes, start at different times of day, and have different running times (lengths).
  - **2)** These paragraphs include variables that will help problem solvers narrow down the movie choices. Encourage students to use text-marking strategies here.
- ● Before attempting the solution, discuss the idea of "Take Your Pet to the Movies Day."
- ● Review the concept of elapsed time and how to calculate it. You might use *Chimps Go Surfing* to help students get started.

Suggested Solution: Based on the criteria mentioned in the text, *Spaniels in Space* is the best choice for Digit. The key points in the story rule out *Tusks and Trunks* (Digit fears elephants), *My Favorite Camel* (the longest film), *Roadrunner's Revenge* (the earliest film), *Band of Bunnies* (ends too late), and *Chimps Go Surfing* (which will get Digit too excited).

Safer Safe (p. 49)

Math Skills/Concepts: number patterns, logical reasoning

Teaching Tips

- ● Begin by discussing with students what makes a safe safe. It is a sturdy container with a door that locks and can be opened only by using a key or a specific combination that consists of an exact sequence of numbers.
- 🔍 **1)** Suggest that students highlight the key detail about the written combination.

2) Encourage students to look carefully and analytically at the list of numbers and on the layout of the keypad. Be sure they understand the uses (given in the text) for the symbols ∗ and #.

☞ The icon here suggests that the solution relies on identifying and extending a number pattern.

● Extend the lesson by talking about the last paragraph in the text. Ask students: *Why is it risky to use a number pattern for a secret combination?* (Smarter thieves could figure out the pattern and therefore the numbers that follow.)

Suggested Solution: The two missing numbers of the combination are 120 and 720. The pattern, based on the first four numbers, is ×2, ×3, ×4, and so on. The full combination would look like this: 1∗2∗6∗24∗120∗720#.

Carpet Confusion (p. 51)

Math Skills/Concepts: geometry (properties of rectangles), measurement (area), visual reasoning, money computation

Teaching Tips

🔍 Tell students that all the data necessary to find the total area of the floor of Camille's tree house is provided in the sketch. Guide them to use their knowledge of the properties of rectangles to determine the length of the unlabeled segment *(7 ft)*.

☞ Ask students to state the challenge Camille faces to answer her mother's question.

● Discuss what is irregular about the floor in the tree house. (Though all angles are right angles, not all sides are equal in length.)

● Talk with students about why Camille's mother asked for an estimate of the cost of carpet squares. Point out that it's common to buy a few extra squares "just in case." Also, explain that there may be added sales tax, and the text describes the price as about $0.60 per square.

Suggested Solution: The total area of the floor is 75 square feet. Find this by breaking apart the irregular shape into two rectangles as follows: (5 ft × 12 ft) + (3 ft × 5 ft) = 75 sq ft OR (5 ft × 8 ft) + (7 ft × 5 ft) = 75 sq ft. To find the cost of carpet squares, multiply: 75 × $0.60 = $45.

Taco Take-Away (p. 53)

Math Skills/Concepts: whole-number operations, multistep problem, factors, multiples, fractions, prime numbers

Teaching Tips

☞ The question about the license-plate number sets the solution process in motion.

🔍 **1)** Most of the details needed to determine the license-plate number appear in this paragraph. Encourage students to mark the text and discuss, as needed, to clarify each hint.

2) This clue provides the final bit of information the detectives need to solve the mystery.

● The crux of this problem is to interpret clues given in mathematical terminology to figure out a 6-digit number. Review the terms *sum*, *product*, and *prime number*, as well as positional terms, such as *middle* and *second-to-last*.

● As needed, clarify the distinction between *digit* (any one of 10 symbols—0, 1, 2, 3, 4, 5, 6, 7, 8, and 9) and *number* (an amount, quantity, or value represented by one or more digits or by words; for example, the number 14 is represented by the digits 1 and 4). Encourage students to use trial-and-error (guess and check) to unravel the 6-digit solution. Remind them that 0 is the least digit and 9 is the greatest.

- Extend by creating similar number problems with clues that depend on interpreting math language.

Suggested Solution: The license-plate number is 4 9 3 3 7 8. To figure out this number, the detectives used the details of each math clue and their understanding of key mathematical terms.

Fair and Squares (p. 55)

Math Skills/Concepts: measurement (area, perimeter), multistep problem, place value (multiplying by powers of 10)

Teaching Tips

- 🔑 This paragraph states the main idea of Little Wiley's problem—how much land he will get from Aunt Molly.
- 🔍 Guide students to read this paragraph closely to understand the process of dividing up the land. Text marking will be helpful, as will listing the various calculations needed.
- This multistep problem lends itself to students working in pairs or small groups because its solution requires a series of calculations done in a particular order.
- As needed, review that area is always given in square units—thus, the correct solution is 4,000 square yards, not 4,000 yards.
- It may help some students to better understand the problem if they sketch a scale drawing of the full parcel of land on grid paper. Suggest that they make a 20 × 20 square (each box = 10 sq yds), and divide it into sections that match the clues in the problem.

Suggested Solution: Little Wiley gets 4,000 square yards. To determine that number, find the total area of the land (200 yds × 200 yds = 40,000 sq yds); then subtract Mom and Dad and Aunt Irene's portions: [40,000 sq yds − (2 × 16,000 sq yds) = 8,000 sq yds]; then divide that by 2 to find Little Wiley's share.

It's a Wrap! (p. 57)

Math Skills/Concepts: spatial reasoning, geometry (properties of rectangular solids), measurement (length, metric system)

Teaching Tips

- 🔑 **1)** In this paragraph, Zara requests the data she needs to solve Marisol's problem. She asks for the three different dimensions of the gift box, which is a rectangular solid (right rectangular prism).
- **2)** This clue suggests a strategy for solving the problem and offers a reminder to add enough ribbon to make the bow on top.
- 🔍 Guide students to carefully study the drawing of the box to understand its properties and dimensions.
- Review the properties of a right rectangular prism: It has six rectangular faces; opposite faces are congruent. Remind students that Marisol plans to wrap the package as shown in the drawing. She will hand-tie the bow on top.

- Extend by providing actual right rectangular prisms (tissue boxes, shipping cartons, etc.) students can measure to determine how much ribbon would be needed to wrap each one and tie a bow, as in this problem.

Suggested Solution: Marisol needs about 360 centimeters of ribbon. To cover all six faces of the box, she'll use (2 × [60 cm + 40 cm]) + (4 × 30 cm) = 320 cm, plus another 40 cm to make the bow, for a total of 360 cm.

Forgettable Flavors (p. 59)

Math Skills/Concepts: logical reasoning, combinations, multiplication

Teaching Tips

- 🔍 A close reading of this paragraph will help students identify the problem that Alex has to solve for her uncle. It also reveals additional information not included in the list of flavors.
- 🔑 **1)** Zara's statement presents the problem succinctly.
   2) Here, Mendel suggests a strategy students can use to solve this type of problem.
- Be aware that some students may instantly recognize that 7 flavors times 3 container choices equals 21 possibilities. Ask them to show how they know.
- Another familiar approach students might use is to create a tree diagram showing all the possible choices. As needed, help students begin such a diagram or make a much simpler tree diagram they can use as a model.
- Extend by challenging students to determine the number of ice-cream flavor/container choices when the customer can order a cup, cone, or bowl with two different scoops.

Suggested Solution: There are 21 possible 1-scoop choices. Simply multiply the number of flavors by the number of containers: 7 × 3 = 21.

Turkey Talk (p. 61)

Math Skills/Concepts: fractions, money computation, multistep problem

Teaching Tips

- Begin with a brief discussion of ways prices are discounted. For instance, they might be advertised as a fraction off the full price ("½ off"), a percent off ("25% off"), or a discount of a set dollar amount ("$1 off"). Provide some examples to focus students' understanding.
- 🔑 Silvio's statement sets the problem in motion—he believes he was overcharged for a recent purchase.
- 🔍 **1)** Discuss the meaning of "⅓ off" in terms of a price. You might offer a simpler example, such as, "What is the price of a $6 hat at ⅓ off?" ($4; ⅓ of 6 = 2; 6 − 2 = 4)
   2) This paragraph presents Silvio's recollection of what happened. Suggest the use of text marking here to highlight key data.
- Extend by creating similar problems (about unit-fraction discounts) for students to solve.

Suggested Solution: The error was subtracting $3 from $15 rather than finding ⅓ of $15 ($5) and subtracting it. Silvio's price should be $15 − $5 = $10.

And the Final Score Was . . . (p. 62)

Math Skills/Concepts: whole-number and fraction operations, number sense, multistep problem

Teaching Tips

🔍 **1)** This bulleted list breaks down each team's score quarter by quarter. Have students highlight key facts, perhaps using different colors for the Fumblers and Dribblers.

2) Refer students to the Basketball Scoring chart, which summarizes the point value for each of three ways to score points in a basketball game.

🔑 Zara's suggestion to fill in a scoreboard like the one shown is an excellent strategy for this type of problem.

● Pair up students or divide the class into small groups. If possible, have at least one student who is familiar with basketball scoring in each pair or group.

● Guide students to fill in each team's scores, quarter by quarter, based on Daeshim's summaries.

● Suggest that students use the "work backward" strategy to determine the one unknown score. *(20 points for Dribblers in the 2nd quarter)*

Suggested Solution: The final score was 79–75 in favor of the Dribblers. The scoring, by quarters is as follows: Fumblers: 17 + 18 + 21 + 19 = 75; Dribblers: 14 + 20 + 24 + 21 = 79

Cat Food Case (p. 64)

Math Skills/Concepts: money computation, multistep problem, fractions

Teaching Tips

🔑 **1)** This paragraph describes the basic problem Lizzie must solve. Discuss with students what *best price* means.

2) Zara's comment hints at how to go about finding the solution. In this case, comparing costs means comparing prices of a 15-ounce can.

🔍 Emphasize the need to read this price list very carefully. Suggest that students use text marking to highlight the different ways the cat food is priced in each of the four stores.

● Explain to students that stores advertise their products and prices in different ways to attract customers. For instance, draw attention to Acme Market's "today only" note and to the deceptive way Cat & Dog Country priced their cat food.

● As needed, review what it means when a price is listed as "$1/5$ off." (Revisit Turkey Talk, p. 61, for a similar pricing style.)

● Extend by comparing prices advertised in newspaper or online ads for local stores. Challenge students to "go shopping" to find the best price for a particular item that would be of interest to them.

Suggested Solution: Best Bargains has the best price—$1.40/can. To find the best buy, Lizzie must determine the per-can price at each store and compare them to find the lowest price.

Hiking Head-Scratcher (p. 66)

Math Skills/Concepts: number sense, logic, multistep problem with fractions/decimals, measurement (distance, time)

Teaching Tips

- This mystery provides an informal introduction to the time, rate, and distance formula.
- Each of the three hand lenses points to a section rich with data that students will need to solve D'Angelo's problem. Suggest that they use text marking, sticky notes, colors, or other methods to highlight key facts. Be sure students note the time the mom dropped off the boys (noon) and the time they need to be picked up (2 p.m.).
- The key to solving this problem is to use the speed the brothers are likely to walk, the lengths of the trails, and the amount of time between when they were dropped off and when they expect to be picked up.
- Clarify the distinction between a loop trail and an in-and-out trail. Explain that on an in-and-out trail hikers walk some distance then turn around and retrace their steps on the same path.
- Guide students to match the length of each of the three trails with the expected pace of hikers along each one to determine when each hike would end.

Suggested Solution: D'Angelo should pick up his brothers at the Lost Lake trailhead because it's the only one that takes about 2 hours to walk. (You might accept the Rock Creek Trail, which can be hiked in 1.5 hours, if students can justify their thinking.)

Party Animals (p. 68)

Math Skills/Concepts: number sense, whole-number operations, algebraic/logical reasoning

Teaching Tips

- To solve this problem, it's essential to know how many animals were at the party and how many feet each type of animal has.
- This is the crux of the problem—knowing how many animals attended the party and how many feet Daeshim counted is the key to finding the answer to his challenge.
- Suggest students use the guess-and-test (trial-and-error) problem-solving strategy. Encourage them to begin with what they know for sure (Vinny the spider is at the party, and a spider has 8 legs) and work from there. Remind them to take each guess and improve it using number sense until they arrive at the correct answer.
- Extend by inviting students to create their own animal-feet problems for classmates to solve.

Suggested Solution: The party had one 8-legged animal, six 4-legged ones, and five 2-legged ones. Start with the total number of feet counted *(42)*, and use number sense and trial-and-error to test different combinations until you land on the one that works.

Relative Ages (p. 69)

Math Skills/Concepts: number sense, logical reasoning

Teaching Tips

🔑 Mischa's problem is that he does not know the ages of his three cousins. Suggest that students read carefully for clues they can use to help solve this problem.

🔍 **1)** Have students use text marking to highlight data in this paragraph about the ages of the cousins. They might use a different color for data about each cousin.

2) Here Mischa reveals the rest of the information needed to solve this problem.

● Have students begin by listing the names of the three Croatian cousins with any straightforward data about them, such as that Ana is the oldest. Ask students which cousin must be the youngest and why. (*Aleksandar; Anton is 3 years older.*)

● Students can solve this problem by applying the given clues, and then making guesses and adjustments until the correct ages are found. Emphasize the importance of knowing the total of the cousins' ages, always keeping that key fact in mind.

● Extend by challenging students to create their own age problems based on the ages of actual or made-up family members.

Suggested Solution: Ana, the oldest, is 16. (16 is 4 squared.) Anton, who is 3 years older than Aleksandar, is 11, which means that Aleksandar is 8. To check, add: 16 + 11 + 8 = 35.

Finders, Keepers (p. 71)

Math Skills/Concepts: multistep problem, operations with fractions with unlike denominators, money, algebraic thinking

Teaching Tips

🔍 **1)** This paragraph provides part of the essential data students need to solve the problem: the fractional portions of the treasure that go to Charmika and Paco. Suggest that they use text marking here.

2) This statement contains information students need to determine hidden information—the total value of the treasure. The first step is to figure out what fraction of the total take is Leo's share.

🔑 The basis of the problem is how Leo will divvy up the money, the total value of which students must determine to answer Charmika's question.

● Help students prioritize the known facts, the unknown information, and how to connect the two. Their first task is to find the total value of the found coins. They know that Leo will keep $25. They know the fractional amount of the total to go to Charmika and Paco: $\frac{1}{4} + \frac{1}{3}$ (or $\frac{7}{12}$) of the total. Guide students to understand that, therefore, the fraction of the total for Leo is $1 - \frac{7}{12} = \frac{5}{12}$. To find that, students must find the number of which $25 is $\frac{5}{12}$. As needed, review finding common denominators.

● There are different approaches to this part of the mystery. Some students will divide $25 by $\frac{5}{12}$ ($\frac{25}{1} \times \frac{12}{5}$) to get the total of $60. Others may break apart $\frac{5}{12}$ into unit fractions: each $\frac{1}{12} = \$5$, so $\frac{5}{12} = \$25$, and $\frac{12}{12} = \$60$, the total value of the treasure. Still others may guess and check.

- Be aware that some students may need support to find the total value of the treasure. You might provide them this information as a hint they can use to figure out how much Charmika and Paco each get.

Suggested Solution: The bag held 240 quarters, for a total value of $60, of which Leo kept $25 for himself. Of the remaining $35, Charmika got $15 (¼ of $60) and Paco got $20 (⅓ of $60).

Tech Time-Crunch (p. 73)

Math Skills/Concepts: whole-number and fraction operations, multistep problem, measurement (time), algebraic reasoning

Teaching Tips

🔍 **1)** This paragraph contains important data students should mark—the time required to complete the project and how much time Ingrid and Felix can devote.

   2) Here students will find the remaining essential information they'll need. Have them read carefully and mark the key data—the number of hours each friend can work.

🗝 The question to answer is how many friends Ingrid and Felix need to ask for help.

- Help students prioritize the known facts, the unknown information, and how to connect the two.
- Ask: *What do cupcakes have to do with this mystery?* (Ingrid mentioned that they might "pay" their friends with cupcakes for helping them out.)

Suggested Solution: Ingrid and Felix need the help of 6 friends. They themselves can put in 11 + 10 = 21 hours, leaving 9 hours of work to be done by others. If each friend gives 1½ hours, the number of friends needed is 9 ÷ 1½ = 6.

Fish Scales (p. 75)

Math Skills/Concepts: multistep problem, operations with fractions, measurement (weight), algebraic thinking

Teaching Tips

🗝 Liane's question to her uncle sets the problem in motion.

🔍 This paragraph contains all the information students need to reach the solution. Have students use text marking to highlight key facts and details.

- Help students prioritize the known facts, the unknown information, and how to connect the two. Be sure they realize the importance of knowing the size of the biggest fish Dave caught. Invite them to explain how they must use that detail.
- Challenge students to explain why 5 ÷ ⅓ yields the same answer as 5 × 3. (Multiplication and division are inverse operations.)

Suggested Solution: Uncle Sid's biggest catch weighs 15 pounds. Half a pound more than 4½ pounds is 5 pounds, and 5 is ⅓ of 15.

Meeting Marvin (p. 76)

Math Skills/Concepts: measurement (time, time zones, elapsed time), multistep problem

Teaching Tips

- Because this mystery contains a great deal of information—essential and not—it's especially helpful to provide copies of the Investigator's Log (p. 25). Using it, along with text marking and ample discussion, may help students work more productively toward the solution.

🔍 **1)** This paragraph is chock-full of information. Encourage students to read closely and carefully and to use text marking to identify the data provided here.

  2) You may want to alert students that this section contains significant information as well as unnecessary data. Guide them to highlight only the data that is pertinent to the solution.

  3) Here's one last bit of necessary data.

  4) The concept of time zones can be challenging to students. Use the given map (or one like it) to discuss this real-world topic. Take a moment to locate your own community in its time zone.

🔑 Here is where readers learn the essence of Ruby's problem—when to leave the house to pick up Marvin from the bus station.

- Students might benefit from working in pairs or small groups to solve this mystery.
- Extend by presenting realistic time-zone problems for travel to or from your location.

Suggested Solution: Ruby should leave home at 4:15 p.m. Students must remember to add 2 hours for the difference between Mountain Time (where Denver is located) and Eastern Time (where New York is located), and subtract 30 minutes for the bus ride.

The Broom King (p. 78)

Math Skills/Concepts: number sense, logical/proportional reasoning, measurement (length, distance), creative thinking

Teaching Tips

🔑 Most students will be familiar with advertising and how companies try to attract customers by making memorable claims. This passage poses that kind of problem for Alex.

🔍 Here, Alex shares her understanding of her open-ended challenge.

- Unlike the other mysteries in this collection, this one has no fixed answer. It's a wide-open challenge to students' creativity, imagination, and number sense.
- Discuss the distinctions between saying, "Broom Closet has sold 52,798 brooms!" or saying, "Broom Closet has sold enough brooms to line the road from Frederick, Maryland, to Baltimore!" Most students will recognize that the first statement may be true, but is fairly dull, whereas the second statement is more visual and piques the imagination.
- Encourage students to work in pairs or teams on this challenge. Invite them to share, explain, and compare their solutions.

Suggested Solution: There are many possible solutions to this problem. But each requires students first to estimate the length of a broom, and then select a sensible benchmark for comparison. For instance, if a broom is about 5 feet long, then 60 of them set end-to-end would equal a distance about the length of a soccer field. So Alex might write something like: *BROOM CLOSET has sold the length of over 150 soccer fields of brooms!*

Name: _____ Date: _____

Investigator's Log

Name of Case:

Restate the problem in your own words:

What information do you have?

What information do you need?

What do you do to solve the problem?

Solution:

How do you know that your solution makes sense?

Meet the Detectives

Get to know our two clever sleuths, their dog, and their consultant.

Zara and Mendel live next door to each other. They are not just next-door neighbors, they're pals. And they are not just pals, they're a team—a team of shrewd math detectives. You see, these two share a keen interest in mysteries—especially the part where they get to solve them.

I could tell you more about the team, but there's somebody who knows them and their work better than I do. Probably better than anyone does. That's because he's generally right there where the action is, where problems appear and then get solved. And by "right there," I mean on the rug or on the floor. This "somebody" is a dog—a big, beautiful, cream-colored, fluffy one. His name is Digit, and he will continue your introduction to Zara and Mendel and what they do.

Digit dictated the following information to me, since, like most dogs, his typing is terrible.

Hi, everyone. My name is Digit. Yes, I am indeed a dog. Often, they say things to me like "Sit," "Stay," or "Come." My favorite, though, is "Good dog!" I am good, really good. And "they" are Zara and Mendel. I belong to Mendel and live with him and his family. Still, Zara sometimes has the pleasure of taking me on walks.

I love my math detectives and admire them for having earned a reputation for excellent problem solving. Actually, they are part-time detectives and problem solvers since they're in fifth grade. They take on cases after doing their homework on school days or in their free time on weekends. And over summer vacations, they work practically full-time. I like helping them a lot, almost as much as I like swimming.

I've told you things that Zara and Mendel say to me. Another is "What a smart dog!" or "What a brilliant boy!" They say those things when I nod my head or wag my tail to agree with them. But they say it with real enthusiasm when I help them or, better still, when I point them in the right direction.

Of course, they say different things when I snatch one of their socks to hide, or when I help finish a snack someone happened to leave on a table for a moment. (Didn't they plan to leave it for me?)

Anyway, let me tell you why Zara and Mendel are a great team and very good at their work. I watch them share ideas, check each other's reasoning, explain things to each other, and figure stuff out together. I've seen them sharpen their math skills by thinking as a team.

I notice that problems come to Zara and Mendel via phone calls, e-mails, texts, letters, and over social media. Sometimes the two may bump into a puzzling predicament while out walking or even at school. They rarely say "no" to a chance to help a confused client or a frantic friend.

When I'm not busy eating, sleeping, sniffing, digging, playing, or scratching, I watch attentively as they employ different problem-solving strategies, use critical-thinking skills, reason logically, and apply number sense. They may even make educated guesses to help clients solve tricky problems.

Yet, they're not perfect problem solvers, so they sometimes need help— maybe just a hint. For that, there's their buddy Daeshim, who lives across the street. Daeshim makes them even better at detecting. He's a little older— in middle school—and is an amazing math whiz. He's their secret weapon. I like him a lot, too, because he usually has treats for me in his pockets. If I haven't mentioned how much I like treats, let me mention it now. Liver treats are my favorite. But salmon or lamb treats will do in a pinch. And I also like lettuce, egg, and green beans.

Thank you, Digit. That was very helpful. I'll take over again.

In the next few weeks, you'll be reading some of Zara and Mendel's recent cases. (Admittedly, Digit helped with some recollections.) Read them carefully, and try your hand at solving them. Maybe you'll come up with a different way to solve some of the problems. Feel free to share your solutions! So, put on your best thinking clothes, clear the room of pesky pets, boot up your brain, and dive in!

Happy sleuthing!

Time and Temperature

When can the brothers get into the pool?

Even though Zara and Mendel stay busy solving math problems throughout the school year, there is no let-up during summer vacation. That's not surprising—summer is when kids often have more time on their hands. More time can mean more problems for the detectives to unravel. It was early one late July morning when Mendel received this text:

Hi, guys. James here. My brother and I have a problem.
It's about swimming. Can you help?

Mendel yawned and repeated the text aloud. There was a problem involving swimming, he mumbled to himself. Digit heard him and thought, "Swimming, water—my favorites!" His tail wagged like the blades of a fan. Mendel appreciated this because it seemed to cool the room.

Mendel yawned again and rubbed the sleep from his eyes. He petted Digit and scratched behind his ears. Then he texted back to James:

Sure thing, James. What's your problem?

The reply came immediately:

We're at Granny's house. We want to swim in her pool, but Granny has a strict swimming rule: Air temp must be at least 80°F. We check the outdoor thermometer constantly.

Mendel then texted this:

When did you check temps this morning?

James texted back in a flash:

Every half hour. Began at 8:00 a.m. It's 9:30 now. Temp is 75°F.

Mendel thought for a moment. These two brothers really wanted to swim, but he needed more information before he could help them take the plunge. He also needed more sleep, but that wasn't going to happen. James had his full attention and Digit's, too. He texted:

What were the temps when you checked?

Digit stared at him, and then barked toward the door. Mendel knew his dog's habits and signals. He quickly texted again:

Taking dog out. Leaving phone. Be right back. Get me those temps.

When Mendel returned, he fed Digit. (A dog will happily eat at any temperature and at any time.) Then he read James's response, which was very helpful:

Checked every half hour beginning at 8. Temp at 8 was 60.
Temp at 8:30 was 66. At 9 it was 71. Now temp is 75.

Mendel displayed the data in a table.

TIME	8:00 a.m.	8:30 a.m.	9:00 a.m.	9:30 a.m.
TEMP	60°F	66°F	71°F	75°F

"Aha!" he said, knowing that he had all the data he needed. He dashed off one final text to James:

Keep checking every half hour, J. If temps rise as they have been, you'll be splashing in that pool well before lunch.

If temperatures continue to change as they have been, at what time will James and his brother get to swim? How did Mendel know?

Dollars and Sense

Which party favors will Buster rustle up?

Sometimes, the two detectives don't get called about math problems; they just bump into them. That's what happened one Saturday when Zara, Mendel, and Digit were strolling in the neighborhood. It was a windy fall day. Digit amused himself by sniffing and rustling in the fallen leaves. Zara and Mendel chatted as they walked. Then it happened.

"Ouch!"

"Wha . . . ?"

"Oops! Sorry. Bumped right into you, Mendel. Didn't see you. I was . . ."

"That's okay, Buster. I didn't see you either. What's going on?"

"Well, Mendel, I'm actually glad we crashed into each other. As it happens, I have a money problem and maybe you can help me make sense of it."

Digit settled into the leaves to lick his paws and listen as Buster presented his problem to Mendel and Zara. It was a shopping puzzle that had him stumped. (That's what he'd been thinking about when he crashed into Mendel.) He explained that his little sister had a small cowgirl party coming up and that his mom had asked his help to pick different favors for each of the 4 guests. She directed him to an online shopping site and told him to spend no more than $5 for all the gifts. Buster wants to come as close to $5 as possible.

"My sister is crazy about westerns. I've narrowed my choices down to six favors. Can you guys help me pick the four favors?"

> **Which four items did the detectives advise Buster to buy?**

Huh?

What became of that bothersome ball?

Kim was nearly in tears when she came running toward Mendel in the park. Mendel was walking Digit at the time and was happy for human company. He liked talking but couldn't get a word out of the dog. He'd been having one of his many one-way conversations.

"Hey, Kim, what's wrong? You seem upset," he said with concern.

"I am upset. It's Daeshim again, and he's got me completely flummoxed."

That was the first time Mendel had heard Kim use the word *flummoxed*. But it wasn't the first time he'd heard about Daeshim being confusing and frustrating.

"What did he do now—insist that a day should have 32 hours?" Mendel asked.

"It's what he said he *did*," she replied. "He told me that in this very park this very morning, he tossed a tennis ball into the air."

Mendel didn't think that was so strange, but tossing the ball into the air wasn't the disturbing part. The part that bothered Kim was that Daeshim not only said that he tossed the ball into the air, but he claimed that he never caught it. Furthermore, he insisted that the ball never landed on the ground.

"How can that be?" Kim asked Mendel. "Is he just teasing me?"

Mendel looked around the park. So did Digit, and both turned to Kim. "You know, Kim, he may not have been teasing you. He may have been challenging your powers of observation. Check out this park and think like a detective. I'll bet you can come up with several reasons why the ball never hit the ground even if Daeshim didn't catch it. I can think of at least ten possibilities right off the bat," Mendel said with a wink as he tossed a ball for Digit to fetch.

> **What possibilities could fit Daeshim's claim?**
> **How many can you come up with?**

Ball Bandit

The detectives use a map to track a missing dog toy.

Every once in a blue moon the detectives tackle a problem that comes from an unexpected source. A dog would be an unexpected source, wouldn't you agree?

Cocoa is a chocolate lab and one of Digit's dog-park pals. Most pooches in the park busy themselves chasing balls or one another. Others spend their time sniffing the weeds along the fence or pestering the humans for treats. But Digit and Cocoa are different. They get bored retrieving tennis balls the humans carelessly toss. They soon have their

fill of play fighting, so the two buddies generally lie side by side in the shade and have a private tail wag over local dog news.

One day some canine news had them buzzing. Their tails fluttered like the wings of hummingbirds. The news was that a new dog in the neighborhood, Pokey, was off to a bad start. But Pokey is anything but sluggish. Pokey is a little greyhound—and a little thief.

Digit and Cocoa learned this from one of their bird buddies. The feathered observer told them that Pokey snatched a ball from Daniel the spaniel and raced off with it. No one could catch him. But the bird saw it all, watching from above as Pokey ran from street to street in the neighborhood. She even saw where Pokey hid the stolen ball.

Daniel the spaniel was very upset. *Dogs do not take other dogs' balls! That is just plain wrong!* Digit informed Cocoa that this was a case for the detectives. He informed his furry friend in his natural but effective dog way of communicating. (Dogs don't say much, but they can usually get their message across to other dogs.) He told Cocoa that he would let Mendel know all about this bad dog business and get him on the case.

Digit used all his canine cleverness to get the crime across to Mendel. He brought him a ball. He barked and whined, he darted back and forth and jumped up and down. He pulled a map off a shelf without knocking anything over and dropped it on Mendel's lap. He even got his leash, gripped it in his mouth, and waited by the door. Digit was a dog on a rescue mission.

You know how only family members can understand what babies are saying when they mumble? Clever dog owners have the same talent—they can figure out what their pets are trying to tell them. While someone else might hear only "Woof, woof," a dog's owner can usually figure out what the dog wants.

Mendel listened carefully and tried to discover clues in all that Digit did. He soon came to understand. It appears that Pokey left the dog park on the northeast corner and raced up Jasmine Street until he reached Dove Drive. There he made a right-angle turn and ran two blocks east. On that corner Pokey took a right-angle turn and ran north for a block before taking a third right-angle turn, this time to the west.

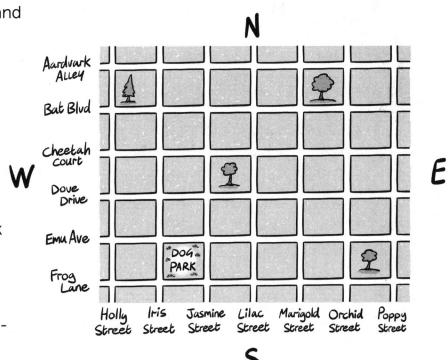

He slowed down along the sidewalk on the south side of the street. There he stopped mid-block and buried Daniel's ball beneath a blueberry bush.

Digit's directions were excellent; the bird has seen and remembered it all. Within half an hour Mendel and Digit returned the stolen ball to a delighted Daniel. Digit proudly carried it in his mouth the whole way.

"Good dog!" announced Mendel enthusiastically.

> **Where did Pokey bury Daniel's ball?**
> **What clues did you use to find the location?**

Too Little Information

Is there enough data to give the right response?

Math detectives like Zara and Mendel sometimes have clients who leave out information. Detecting is a lot harder when there is too little information. Kenta was one such client. He called Zara's cell phone. "My neighbor just asked me a question I couldn't answer, so I called you," he said. "My problem is about mowing lawns.

"I was mowing Mr. Weedler's lawn when his neighbor, Ms. Greengrass, looked over her fence and waved to me," he continued. "I stopped the mower and went over to see what she wanted. She wondered whether I would be interested in mowing her lawn as well. And then she asked what Mr. Weedler pays me."

"How did you respond?" Zara asked.

"I told her that Mr. Weedler pays me $20. The lawn around his corner house is 100 feet wide and 60 feet from front to back. I estimated that I'd be done with his lawn in about 45 minutes, after which I could take a short break and begin mowing hers.

"She was delighted," Kenta continued, "but wondered how much to pay me."

"And you couldn't answer?" Zara replied.

"Right," Kenta answered. "I was stuck, so that's where you come in."

"Hmm," Zara said. She asked him to hold for a moment while she discussed the issue with Mendel. In a moment she got back to Kenta. "Thanks for your confidence in me, but there is something more you need to tell me. Without it, I can't help you."

What information is missing? How might Zara and Mendel help Kenta answer Ms. Greengrass?

Too Much Information

The detectives ignore extra information to nail down a solution.

Having too little information can stop Zara and Mendel in their tracks, preventing them from finding a solution. But sometimes clients tell them far more than they need to know. This also can be a problem, because too much information can clog a case and slow down even the best detectives.

"Extra information can get in the way. Be on the lookout for stuff you don't need to know," Daeshim once advised them.

"Some people don't want to leave anything out," Zara noted. "They tell us more than we need to know, sometimes way more."

That's exactly what recently happened to our detectives during a phone call. Mendel was on one end of the line. The person on the other end was Ruth, calling with a problem about hammering nails.

Ruth phoned from the mountain lodge her family owned. The season was over, so Ruth's family was at work closing up the property. One key job was to protect the cabins from winter snow damage. Ruth's task, with her brother's help, was to hammer nails into boards to shutter the windows. There were 10 cabins; each had 1 door and 2 windows.

Ruth said that each window needed 6 boards, and that each board needed 4 nails. While her brother held each board in place, she hammered. "I can knock a nail all the way in in just 30 seconds," she announced proudly. "My problem is that my dad asked me when I would be done. He has another job for me. But I'm stuck," Ruth admitted, "because I don't know what to say."

"How far along are you in your work?" Mendel asked.

"I'm on the last cabin. Just 1 more window to go!" she answered. With that last detail, Mendel had Ruth's problem solved.

> **What did Mendel tell Ruth to tell her dad?**
> **What information did Ruth give that was not needed?**

Doggy Dilemma

What do you do when dogs can't get along?

Zara and Mendel often do their homework together. That's what they were doing after dinner one Monday evening when Zara's cell phone rang. Digit was there, too—curled on the rug, dozing. Asleep no longer, he sat right up and listened. He always liked to know what was going on.

"It's Luis Gomez," Zara told Mendel.

They know him and so does Digit. Luis has that perky poodle, Lola. (Pesky poodle is more like it.) Maybe he's about to give Zara and Mendel a new problem to solve. Digit hoped so because watching them do their homework puts him to sleep. Well, actually, a lot of things put him to sleep. But he loves to watch them solve problems. He'll stay wide awake and give them his full attention.

"I'll bet he has a problem for us to solve, Zara," Mendel replied. "And I'll wager it's got something to do with that dog of his, that Lola."

It did have to do with Lola. Luis told Zara that he had taken Lola to the dog park that afternoon. He said that Lola loves going and he loves taking her. But when he spotted Jamie coming into the park with Jasper, he acted. He hurriedly put the leash back on Lola and hustled her out. "You see," Luis explained, "Jamie and I are pals, but Lola and Jasper simply can't stand each other. When those two pooches are together, the fur flies. Jasper howled when he saw Lola. Lola barked on her way out."

Luis explained that he takes Lola to the dog park *every other* afternoon. "It's important to stick to Lola's schedule," he insisted. He then told Zara he knows that Jamie brings Jasper to that same park *every third* afternoon. Since both dogs are on a schedule, when they go to the park on weekends, it's also in the afternoon. And he made it clear that Lola and Jasper cannot be in the park at the same time. That's the dilemma.

Then Luis got more specific. He told Zara that his cousins will be visiting next weekend and he wants to take them to watch Lola play in the dog park one of those two afternoons. He wants them to see how she romps, fetches, chases, and plays tug and keep-away. But he must know whether he can safely bring his dog and his cousins to the dog park one afternoon over the weekend. It must be a day when Jasper won't be there.

Zara listened carefully, as she always does. She told Luis that she would consult with Mendel and then get right back to him. And that's what she did, because she and Mendel believe that two heads are better than one.

After a brief chat, Zara had a solution. "Do you have a calendar handy, Luis?" she asked. "Or can you make a chart showing the days between now and next weekend?"

Can Luis safely bring Lola and his cousins to that dog park one day next weekend? If so, which day will it be? If not, why?

Puzzling Payment Plan

A dog walker faces a confusing decision.

Della loves dogs. She's a dog walker, a dog sitter, and a pal of Zara and Mendel. They were glad to see her when she dropped by with Floppy. So was Digit—and his tail. It spun like the whirling blades of a helicopter; he nearly took off!

But Della didn't just drop by for a neighborly visit. She had a dog-walking problem that needed a solution. The detectives were all ears. Digit showed his interest by jumping and twisting. Soon he grabbed a bone to chomp on to calm himself down. Then, he and Floppy happily sniffed each other. But he was all ears, too, because he liked nothing better than a dog problem— except eating and swimming, that is.

While Floppy and Digit played and sniffed, Della talked to the detectives. She told them about her new client, Mr. Barkley, and his two dogs. Both were golden doodles. Inez weighed 70 pounds, and Izak weighed 75 pounds. Della really looked forward to getting to know these big, fluffy pooches. Mr. Barkley wanted to hire her to walk his dogs together once each afternoon. He told her that he could walk them himself before work

in the mornings and after dinner in the evenings. But he needed someone for the afternoon walks because his regular dog walker was going away to visit her family in Alaska. He told Della that he needed her for 8 days.

Della told Zara and Mendel and Digit and Floppy that her problem was about payment for that 8-day job. How she was paid was not ordinarily a problem. Della usually charged $20 to walk a pair of dogs. Each walk would last for about an hour. But Mr. Barkley made her an interesting offer unlike any other she'd had.

"He gave me two choices to consider," she told Zara and Mendel and Digit and Floppy. Then she explained the choices. "He offered to pay $20 for each walk OR pay me $1 for the first walk and double the amount for each walk after that. How do I choose?"

Zara and Mendel agreed that this was indeed a puzzling puppy problem. Digit and Floppy agreed that the walks would be even better if treats were doubled after each one.

"So which plan do you think I should take?" Della asked. "Which pays better?"

The two detectives huddled for a moment. It was Zara who explained their thinking. "You can solve this problem on your own, Della," she said. "Just make and complete two tables. Then you'll be able to see for yourself which choice to make."

"Make tables like these," added Mendel, showing Della his sketches. "Finish the tables, then compare the totals after the 8 days. The answer may surprise you!"

Table 1: Pay Stays the Same Each Day = $160 ✗

WALK	1	2	3	4	5	6	7	8
PAY	$20	$20	$20	$20	$20	$20	$20	$20

Table 2: Pay Doubles Every Day = $272 ✓

WALK	1	2	3	4	5	6	7	8
PAY	$1	$2	$4	$8	$16	$32	$64	$128

Which plan should Della take?
Explain how using tables helped her decide.

Code Crackers

A familiar child's game shapes a solution.

Any good detective needs to be able to crack codes, as Zara and Mendel know very well. Clients expect it, and crooks do, too.

Daeshim's friend Ollie is not really a crook. Mostly, he's a prankster. He likes making silly jokes and rearranging things in unexpected ways. He likes hiding pencils and pens or using them to change grades from Bs to Ds. He enjoys waking a sleeping friend by putting whipped cream on his hand and tickling his nose with a feather. He'll even play tricks on dogs, as Digit can confirm.

Ollie's latest prank came to Zara in the mail in the form of a brief letter. When she opened it, she saw that its words were formed by letters cut from restaurant take-out menus and pasted onto a sheet of paper. It wasn't signed, nor was there a return address.

She brought the letter to Mendel to discuss the strange message. It plainly stated that Digit's favorite bone was now buried somewhere in the backyard of Daeshim's house. (That probably explained why the dog had been sulking under the kitchen table all day.)

But that was not all. There was something else inside the envelope— a second sheet of paper. Zara took it out and shared it with Mendel. It was a code made with simple shapes, like this:

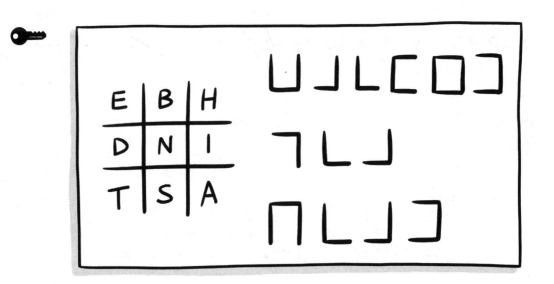

"This must be from Ollie," they said at the same time. Digit growled at the sound of that prankster's name.

Zara and Mendel examined the shapes that made up the code. They looked it over every way they could think of. They looked at it right-side-up, upside-down, right to left, left to right, and even sideways.

"Daeshim's house has a pretty big backyard," Mendel noted with a sigh.

"It sure does," added Zara, frowning. "There's a lawn and a flower bed and even a small greenhouse. And there's also that small building where they keep equipment like rakes, shovels, lawn mowers, and leaf blowers. If we have any hope of finding Digit's beloved bone, we'd better crack this code."

"We may need a brain boost. Should we take this to Daeshim?" Mendel suggested.

But Zara's frown began to change, first into a look of deep thought. Then it became a grin that widened into a full-blown smile.

"What's the big smile for?" Mendel asked. Digit then emerged from under the table, also wondering.

"I got it!" Zara shouted. "I cracked the code! You play tic-tac-toe, don't you?"

"Of course, I do. Everyone does." Mendel paused for a moment, deep in thought. He looked again at the code. Then a light bulb went on over his head. "Wait a minute—I've got it, too!"

"Come on, Digit! Let's go dig up your bone!" the detectives sang out. Digit responded by excitedly chasing his tail.

What have the detectives figured out?
Where is Digit's bone buried?

Musical Chairs

Zara and Mendel help set up
an orderly outcome.

Ms. Basso is the music teacher at Zara and Mendel's school. She also directs school plays and special performances. She is known to be very fussy; she is very clear about what she expects of her actors, singers, dancers, and musicians. She even has high expectations for the stagehands. Kevin is a stagehand who came to the detectives with a problem—a seating problem. And it was Ms. Basso's fault, or so Kevin claimed.

Kevin was in charge of setting up rows of seats in the school's performance space. He was struggling to understand how to follow the specific instructions from Ms. Basso. "She wrote out a seating plan for me to follow. You should see this thing! She's got to be the fussiest teacher in the school," Kevin complained.

Both Zara and Mendel knew this from the one play they had performed in. They agreed that fussiness and Ms. Basso went together like toast and jam. But they also agreed that she was terrific at what she did. After all, the school performances were always memorable.

"Listen and learn," Zara's mother had advised her when she was complaining. Zara thought that was decent advice, but it didn't make her job as set designer any easier. But that was then, and today Kevin is the one struggling with Ms. Basso's demands.

She and Mendel were ready to help. "So what's your trouble, Kev?" Mendel asked.

"Here's the deal," he answered. "There are 200 folding chairs stored beneath the stage. Ms. Basso demands 15 rows, *exactly* 15. So far, so good, I thought. But then she added some rules, but they don't make much sense to me."

Kevin continued his complaint. He told the detectives that Ms. Basso wanted 3 rows up front with 10 chairs per row. Then she wanted the next 8 rows to have 12 chairs each. Lastly, she insisted that all remaining rows must have 15 chairs.

As Kevin described his problem, Zara and Mendel were doing mental math. They looked at one another until Zara spoke. "There will be some chairs left over, Kev. What's to be done with them?" she asked.

"I'm supposed to place the leftover chairs along the walls on the left and right sides of the room. She wants the same number of chairs on both sides. Yikes!"

"Well, you've got some figuring to do, Kev," Mendel added.

"Don't I know it!" Kevin replied. "My head is spinning, and I don't even know where to start."

Zara assured Kevin that he had all the information he needed to arrange the chairs as Ms. Basso wanted. "It's always a good idea to start with what you know, and work from there." But she added that what he really needed were helpers. "Unfolding and setting up 200 chairs is a big job, no matter how you arrange them!" she laughed.

"Yeah, I know that, too," Kevin sighed.

> **How many rows will have 15 chairs?**
> **How many chairs altogether will be in rows?**
> **How many leftover chairs will go along each wall?**

Suki's Salamander

Suki wants to build an amphibian enclosure.

Suki's animal kingdom has been growing. First, there was a mouse. When it ran away, she didn't fret; she simply got herself a salamander. She loved that creature so much that she gave it her favorite name: Sally.

"Nice salamander," Zara remarked when Suki came over with her new pet.

"Nice little salamander, yes. But I have a *big* problem."

Zara wondered how you could even have a problem with a salamander. "What's up, Suki?" she asked.

"It's about the pen I'm designing for Sally to scoot about in. I plan to drop an ant or two in there, maybe even the occasional spider or worm."

Hearing Suki's voice, Mendel came into the room for entertainment. "So your problem is how to keep the food from escaping?" he asked.

"No, smarty pants, my problem has to do with the shape of the pen. I've already got 84 square inches of popsicle sticks to use for the floor. Here's the real problem—Sally is long and lean, and needs at least 2½ inches to turn around in. So I want to make the longest, leanest rectangular pen I can with the flooring I have. That way she—I think she's a girl salamander— can really zip along."

Zara and Mendel knew at once that factors would factor into the solution to Suki's pen problem. And they knew exactly the right size to make it, given the popsicle-stick floor.

> **What measurements did Zara and Mendel tell Suki to use for the length and width of Sally's pen? How did they use factors to figure that out?**

Fountain of Wishes

School projects get funding from a fountain.

Cleo's class had a problem to solve. She didn't waste any time—she brought it directly to the math detectives. Zara and Mendel were glad to take it on since they go to the same school. Digit was glad, too, because the problem involved water.

It actually wasn't so much about water as it was about what was in the water.

The kids attend Einstein Elementary School. Einstein prides itself on how it works with the community to make itself a better school. It always looks for ways to make learning better and more fun. And it always encourages students to take part in their own education.

But making improvements takes time and can be costly. So Einstein School often undertakes projects to raise money—bake sales, art sales, book sales, and old stuff sales. Another way they raise money is to use a fountain.

The fountain they use is located by the front entrance to the school. It's a round fountain, not too big, not too deep. The water trickles down from what looks like an owl's beak. Digit always wants to jump in whenever they pass by. But Zara and Mendel don't let him.

Pennies, nickels, dimes, and quarters line the bottom of the fountain. The school janitor collects them each week and puts them into a large can. The coins are used to buy things the school needs. They might go toward fixing a broken seesaw in the school yard, replacing worn library books, or providing refreshments at school parties.

People like to toss coins into the Einstein fountain. Parents, grandparents, teachers, students, former students, babysitters, dog walkers, and passersby from the neighborhood all contribute. They gladly flick coins into the water as they make a wish.

Once a month, a different class gets the job of computing the total value of all the coins in the can. This month the task has fallen to Cleo's class.

"This month's coins will go toward buying grow lamps for the science lab and e-readers for the library," Cleo told Zara and Mendel. "I was a coin sorter. Other kids counted each kind of coin, after which others recounted just to be sure. We got a grand total of 320 pennies, 96 nickels, 90 dimes, and 28 quarters."

The detectives thought this was a pretty good chunk of change.

"Yes, it's been one of our better months," Cleo said. "My problem is not about how much money we counted. It's about how to spend it." She explained to the detectives the plan her teacher wanted to follow for how to spend the money. He said that for every dollar that goes to buy grow lamps, twice as much should go to purchase e-readers. But Cleo and her classmates weren't sure how to determine how much money each item should get.

"It may take a few more steps, Cleo," Mendel said, "but we can help you solve this."

**How much money goes toward grow lamps?
How much goes toward e-readers?
How did you figure it out?**

Digit Goes to the Movies

Which picture will they pick for the perky pup?

Sometimes Zara and Mendel find themselves solving problems of their own, such as what happened recently. That unexpected problem had to do with Digit.

Digit was blue, so he spent an entire day moping under the kitchen table. And he wasn't there because the kitchen was the magic food room, with all its delicious goodies. No, he was sad for another reason—his next-door neighbors had just moved away. That meant that Pebbles had just moved, too. Pebbles was Digit's pal, even though she was a cat.

When Digit is sad, Mendel and Zara are sad also. So Zara spoke with Marisol, the dog walker, who usually has helpful ideas about dog dilemmas.

"Take him to the movies!" Marisol suggested.

"The movies?" Zara asked. "That's an odd suggestion."

"Not that odd," Marisol replied with a knowing smile. "Tomorrow afternoon is 'Take Your Pet to the Movies Day' at the six-plex. Every movie they show will be about animals, and every pet owner can bring one animal for free."

Zara knew a good idea when she heard one, so she shared it with Mendel. He agreed that the idea could work, but he reminded Zara that there are many things to think about before choosing a movie and a movie time. He pointed out that, like all dogs, Digit is a creature of habit. He does things the same way at the same time every day. There are set times for his walks, his play, his rest, and his meals.

Zara agreed, so they looked at the next day's movie schedule together. Digit was listening the whole time. Something about the conversation made him think happy dog thoughts. He crawled out from under the table, but just partway.

Here's the schedule Zara and Mendel found for "Take Your Pet to the Movies Day."

Movie	Starting time		End time
Chimps Go Surfing	1:40	20 min	3:00
Spaniels in Space	1:50	50 min	3:40
Tusks and Trunks	1:25	40 min	3:05
My Favorite Camel	2:10	55 min	4:05
Roadrunner's Revenge	1:20	75 min	2:35
Band of Bunnies	2:45	2 hours	4:25

"Well, I see we have some decisions to make," Mendel said. As they studied the choices, the detectives considered what was best for Digit.

"We can't take him to the earliest movie because it interferes with his walk," Zara noted. "And Digit wouldn't be able to sit through the longest movie."

Mendel agreed. "And I know that he's afraid of elephants. He thinks they are the biggest, scariest dogs on earth, so elephant movies are out."

"And Digit must be home by 4:30 because that's the time for his play date with Jolly the collie. It's a 10-minute walk from the theater to Jolly's house," Zara added.

"Right. And we shouldn't take him to a movie that features water. You know how excited he gets about swimming," Mendel mentioned wisely.

"Ahhh, I see the movie we should take him to," Zara said. Mendel looked again at the schedule and winked in agreement.

Which movie did Zara and Mendel pick for Digit?

Safer Safe

Not all safes are completely safe!

Not all thieves are good at what they do, which is good news. And some safes are better than others. That's good news, too. Both bits of good news figure in the problem Jolene had; both explain why she called Zara and Mendel.

"The safe in our den is amazing!" Jolene told Mendel and Zara as she arrived at Zara's house. "But we're having an unexpected problem with it."

"That's terrific but unfortunate," Zara commented, wondering how a safe could be both.

Jolene then explained the safe snag. She said that the night before, the family was out to dinner. In their absence, a thief broke in and tried to open the safe. She said they knew this because they saw a candy-bar wrapper on the den floor and messy chocolate fingerprints all over the safe's digital keypad.

"Luckily, our robber flunked at the job," Jolene added. "He successfully slipped in through an open window, but failed to figure out the combination to open the safe."

Mendel pointed out that the thief might have been a "she" and not a "he." Digit, who was listening as usual, knew for certain that it wasn't one of his dog pals. He knew this because none of them have thumbs and most of them can't open anything.

"Either way, I think the failure was good news," Zara concluded.

"The news is actually too good," Jolene admitted. "That's our problem. The safe works too well. None of us at home can remember the combination—so we can't open our own safe!"

Zara then asked the obvious question. "You mean that you don't have the combination written down somewhere, er, safe?"

Jolene answered sheepishly, "We did once. But Lulu chewed the slip of paper it was on. The last two numbers of the 6-number combination are completely gone."

Digit had a hurt look on his face, thinking that perhaps a dog was to blame. He felt much better once he learned that Lulu was a cat.

Jolene continued. "Since Lulu is my pet, it's my job to come to you guys for help. So here are the first four numbers."

Jolene took out the slip of paper on which she'd written those numbers. She'd also drawn a picture of the keypad. She placed her drawing on a table, explaining that the * is used to separate each number from the next and that the # shows that the combination is fully entered.

The detectives studied the numbers, looking for relationships among them. Within moments they felt certain that they had found the solution to Jolene's problem. "Look carefully at those numbers," Mendel guided her. "We believe there's a number pattern in there, something clever your family must have planned."

"It stumped the messy burglar, but you can figure it out," encouraged Zara.

But the detectives suggested that using a number pattern when a more secure set of numbers is needed may be foolish. After all, there might be smarter thieves out there.

> **What are the two missing numbers needed to open the safe? How did you figure them out? Write out the full combination that opens the safe.**

Carpet Confusion

Camille completely carpets the floor of a tree house.

Neither Zara nor Mendel knows much about interior decorating. They never much cared about furniture, rugs, lamps, wallpaper, and stuff like that. Neither could tell you the difference between a love seat and a sofa. Neither could tell you what colors go best with purple or where lamps should go to give a room its best light. And both would have to search the internet to tell you what an ottoman is. (It's a low, cushioned seat without arms or a back.) So when Camille approached them with a decorating problem, they doubted they could be of much help.

"Oh, but it's really more of a math problem," she said when Mendel expressed his doubts. "A tree house–decorating math problem, that is."

Now she had their attention. Not every day did a tree-house problem come their way. For them, the main problem with tree houses is that neither of them had one.

"How can we help you, Camille?" Zara asked.

Camille reached into her pocket for the drawing below. She explained that it was the floor plan of the multi-tree structure her dad was building for her and her brother. She went over it with the detectives, noting its irregular shape and providing the lengths of most of its sides.

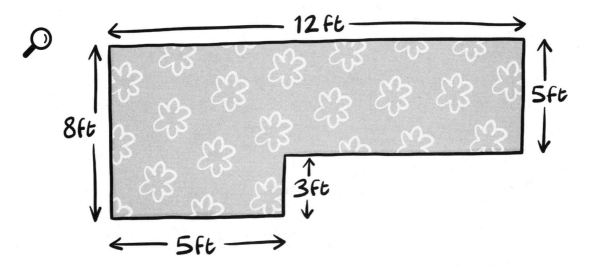

"That's some tree house you're going to have, Camille," Mendel remarked. "It would surely make a great office for Zara and me. I'd put a desk along the long wall and maybe a refrigerator against one of the shorter ones. And then I'd . . ."

"Uh, Mendel, it's not a furniture problem," Camille interrupted. "My problem is about carpeting. My mom knows that a wooden floor gets cold." Camille explained that her mother offered to carpet the entire floor of the tree house, thinking that carpet tiles would do the trick and keep the cost down.

"My mom wants to estimate how much the carpeting will cost. She knows that carpet squares sell for about $0.60 each, and that each carpet square covers 1 square foot. So I guess I need to find the area of the floor. Can you help me with this?"

Zara responded, "Sure we can help. Do you know how to find the area of an irregular figure? For this is precisely what we need to do to answer your mom's question."

Camille said, "Well, I know how to find the area of the rectangle."

"Then you already know what to do. Let's look at the floor plan together, keeping rectangles in mind," said Mendel. "And when do you think you might have your tree house–warming party?"

> **What is the total area of the floor of Camille's tree house? How much should her mom expect to pay to carpet it?**

Taco Take-Away

How can you capture a runaway taco taker?

One thing Zara and Mendel have in common with their genius pal Daeshim is that they all thrive on solving problems. They especially enjoy solving tricky math problems. Another thing they have in common is that they all love to eat.

One day the three friends were walking on Fourth Street. Fourth Street is known for its food trucks. Digit was with them because, like most dogs, eating is one of his favorite activities, too.

"Let's see what looks good today, guys," Daeshim suggested. Digit thought that was a great idea—because food trucks mean tasty treats sliding out of tacos. They mean beef and chicken slipping from buns, or pizza toppings tipping off a steaming slice. Digit thinks of the ground as his doggy snack table.

They had stopped by a truck selling Indian food when they noticed some excitement across the street. A woman on a motorcycle was speeding away from a Mexican food truck. She had just grabbed a taco from an unsuspecting customer and was roaring away, laughing and eating.

But this was no laughing matter. It was a robbery!

Zara and Mendel missed most of the action. They were reading the menu, trying to decide between the lamb curry and the coconut rice. Digit missed it as well. He had been halfway under the truck, snatching a dropped piece of chicken.

But Daeshim saw the whole thing. "I'll bet it was a fish and mushroom taco," he said. "I can tell by her smile." Daeshim not only has a nose for numbers, he has eyes like a hawk.

"I'll take your word for it, Daeshim. But did you see anything else? Did you catch the license-plate number on the motorcycle?" Zara asked. Math detectives do take an interest in actual crimes, like this one. They look for clues right away.

Of course, he noted the license number. But because he was who he was, Daeshim did not make it easy on Zara and Mendel.

"I did see the plate. I saw it with my hawk-like eyes." Daeshim explained what he saw to the detectives and to a police officer who had just arrived. The math whiz said that the license plate had six digits. He noticed that the first digit had half the value of the last, and that the sum of those two digits was 12. He revealed that the middle digits were the same. Daeshim noted that the second digit was the product of the two middle ones. It was also the greatest of the six digits. Then he remarked that the second-to-last digit was a prime number greater than 5.

"Is that all?" asked Zara.

"Are you always this annoying?" asked the police officer.

"Yes!" answered Zara and Mendel together.

"Woof!" barked Digit.

"Oh, and all six digits add to 34," Daeshim added. "Does that help?"

It certainly did. Zara and Mendel used Daeshim's clues to figure out the correct license-plate number. They told it to the police officer, who will use this detail to help capture the crook.

"Now, let's eat!" Daeshim exclaimed. Digit panted in agreement.

What was the license-plate number of the motorcycle? How did Zara and Mendel use the clues to figure it out?

Fair and Squares

Lots of land lands in Little Wiley's lap.

Aunt Molly is moving to Florida. But the land she owns in Minnesota is staying right where it is. And it will stay in the family, too.

That's what Zara and Mendel heard from Little Wiley, who lives down the street from them. He lives there with his mom, his dad—Big Wiley—his cat, his hamster, his turtle, and his sister, Lucy. He'd just gotten some interesting news from his mother and father. They'd heard it from Aunt Molly—Big Wiley's older sister.

"We're going to own some land of our own in Minnesota, Mendel. All of us are!" Little Wiley said proudly.

"That's great!" Zara said.

"That's pretty cool, Little Wiley," Mendel agreed.

"It's only cool there in fall and maybe spring. But in the winter, it gets wicked freezing up there. And in the summer it heats up, according to Aunt Molly."

Mendel ignored those details. He recognized that Little Wiley was super smart and knew tons of facts. But Little Wiley often took a different meaning than intended for words that had several meanings.

"Okay, so here's my problem," Little Wiley continued. "I don't know how much land I'll own for myself. I want to set up a turtle and hamster habitat, but it would help to know how big I can make it."

Mendel smiled. "What do you know about how your aunt plans to divide the land? Can you shed any light on that?" he asked.

Mendel was relieved when Little Wiley didn't start explaining how light it was in his tool shed or how brightly the sun was shining. Instead, he simply answered the question.

"Here's what I know," Little Wiley announced. "The land is a huge square that measures 200 yards on a side. Aunt Molly will give Mom and Dad a section with an area of 16,000 square yards. She'll give Dad's other sister—Aunt Irene—a section the same size. The rest of the land gets split evenly between me and Lucy. Aunt Molly isn't giving any land to my cat, my hamster, or my turtle, and that's a shame."

Mendel recorded the data and did some calculations. When he finished, he told Little Wiley that the land he would get was not only large enough for hamster and turtle habitats, it was big enough for something else.

"Yeah, like a sizable cat play area?" Zara suggested.

"Yessss!" Little Wiley yelped in delight.

> **How much land is Aunt Molly giving to Little Wiley?**

It's a Wrap!

Marisol wraps a gift with ribbon and a bow.

Marisol had a problem with gifts—not about choosing or remembering to send them in time, but about wrapping them with ribbon. If Digit could talk, he would say, "Get lots of ribbon and use whatever you need. Then run around with the rest in your mouth, drop it somewhere for someone else to pick up and put away, and take a nice nap."

But that answer wouldn't have worked for Marisol. The detectives knew they needed to come up with a solution better suited to humans.

"My dad gave me a job, and it's got my brain tied in knots," she told Zara and Mendel as they listened on speakerphone. (A speakerphone is a great tool for a detective team.) "We bought some sweaters for Aunt Eleanor for her birthday," Marisol continued. "She's turning 75 years old next week."

"That's wonderful. I'm sure she will love your gift and good wishes," Zara said. "Now what about that job you mentioned?"

"Ah, yes. The shop clerk put the sweaters in a gift box. I'm supposed to wrap a ribbon around it and tie a bow. I'll use lavender because that's Aunt Eleanor's favorite color."

Hearing that reminded Mendel of his birthday, which was coming up. What gifts might he get? He was hoping for a new skateboard, the kind with a motor. He began to picture himself speeding along the sidewalks while everyone stopped and stared at him in wonder.

"Who's saying 'zoom, zoom'?" Marisol asked.

But Zara stayed focused. "Not me. Now, how big is this box, Marisol? Please tell me its length, width, and height."

"Okay. I'll get a tape measure and the box and take all those measurements," Marisol said.

Zara and Mendel waited while Marisol fetched a tape measure and gathered the data. In the meantime, Zara got out a pencil and some paper. She drew a box with the shape that a box holding sweaters would have. When Marisol came back to the phone, she said each measurement aloud.

Zara labeled the box she had drawn with those measurements. She also sketched where the ribbon should go—right down the middle of each face of the box, like this:

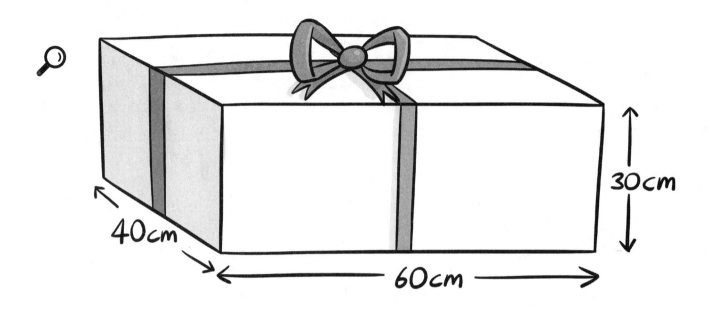

"I'll bet you can figure this out yourself, Marisol," Zara said, "because the box is a kind of figure you know. Think about how many faces it has. Use your knowledge of its properties. That will guide you to estimate how much ribbon you'll need to wrap it completely. Drawing a sketch of the box with its sides labeled with their measurements will help, too. That's what I just did. But remember to add maybe 40 centimeters for the bow."

"Yeah, bows are nice," added Mendel absentmindedly. "Don't forget the bow." He was picturing how his birthday skateboard would look wrapped with a bow.

At least how much ribbon does Marisol need to wrap the box and tie a bow on top?

Forgettable Flavors

How many appalling ice-cream flavors can there be?

Every so often, the detectives get a client who brings them some real food for thought. That happened recently when Alex arrived after her first day at work at her weekend job. She worked for her Uncle Max, who owned an unusual ice-cream shop. Even its name was weird: I SCREAM FOR ICE CREAM.

"It's the oddest ice-cream shop in the world," Alex told Zara and Mendel, who wondered what could be odd about an ice-cream shop. The answer came soon enough.

"The flavors are nuts," Alex said, rolling her eyes.

The detectives wondered whether she meant that all flavors included nuts. They imagined flavors such as butter pecan or vanilla almond. But they quickly got the true picture.

"My uncle hired Monkey Business Marketing to tell him what flavors people say they really want to eat. You won't believe what their research told them!"

"Try us," Zara said with a smile. This was getting interesting.

Alex pulled out this list from her pocket.

Blueberry mustard

Sweet potato & chili

Peanut butter & olive

Corn-on-the-cob

Stinky cheese & mint

Grapefruit & avocado

"For goodness' sake—who would eat flavors like that?" Alex asked. Mendel and Zara knew that Digit probably would.

"Okay, Alex, we get it. But why did you come here for help? Did you think we could convince your uncle that Monkey Business Marketing has gone bananas?" Mendel asked with a wink.

"No, we're stuck with those flavors, at least for a while. My uncle's problem is that nobody will eat any of those gross flavors. My problem is that my uncle wants me to make an appealing sign listing the flavors. And he wants me to mention the number of possible 1-scoop ice-cream choices I SCREAM FOR ICE CREAM customers can select. We offer all the flavors on the list plus a brand-new one—fig and grasshopper. Fig and grasshopper—good grief! But there's more. We sell ice cream in cones, cups, or bowls."

"So you're not sure how to figure out the total number of disgusting choices?" Zara asked.

"Yup," Alex answered. "It all makes my stomach churn."

Mendel and Zara huddled for a moment before Mendel spoke. "If you can make an organized list, Alex, you can solve your advertising problem," he said.

"You might even solve it before you've finished your list," Zara added.

Alex said she would try this idea and thanked the detectives for their help. As she left, she invited them to come by the shop for a free ice-cream cone on Saturday. "The store opens at 11:00 a.m. and closes at 8:00 p.m. I suggest that you come either before 11 or after 8. You'll be a lot happier that way."

Everyone but Digit laughed and agreed.

How many 1-scoop ice cream + container choices are there? How can Alex figure that out?

Turkey Talk

The detectives help Silvio pay the right price.

"I got robbed, I think."

Those were the first words out of Silvio's mouth when he noticed Mendel and his dad at the farmer's market. Holding a bag of turkey sausages, he made a beeline to Mendel the moment he spotted him. It was common knowledge that Mendel was a math detective.

"How were you robbed?" Mendel asked.

"Quickly and easily by that greedy grocer who sells terrific turkey products!"

"Okay, it happened quickly. But what, exactly, did he do?" Mendel responded.

"He charged me too much for this delicious turkey sausage," Silvio replied, holding up his purchase and frowning. "There's the sign I saw by his stand, and I believed him."

ALL PRODUCTS ⅓ OFF TODAY!

Mendel understood that Silvio was upset. So he decided to hear him out despite the fact that he didn't like turkey sausage even a little. "How much did you pay for the turkey sausage?" he asked.

Silvio replied that he paid $12 for the sausage but that the regular price was $15. He told him exactly what the grocer said. "He said, '$15 minus a third—$3—for a total of $12. So I handed over three $5 bills and he gave me $3 back."

Mendel looked at Silvio and smiled. "Come on, Silvio. Let's go see that turkey guy and straighten this out. The error is his, so let's go explain it to him and fix this."

> **What error did Mendel notice?**
> **How much should Silvio's sausage cost?**

And the Final Score Was...

The Fumblers and Dribblers play another close game.

Daeshim often makes solving mysteries easier for the detectives. He helps them sift through clues and points them in the right direction when they feel lost. He even teaches them some useful math skills when necessary.

But sometimes, he is part of the problem. An example of this happened recently to Nicole.

"That Daeshim is driving me crazy," Nicole told Zara when she came over to her house. Mendel was there, too. They were all doing homework together as they often do. Digit was there as well, working on one of his favorite bones. He often does that, too.

What Daeshim drove Nicole nuts about was a basketball game between the Fumblers of Jordan High and the Dribblers of Jabbar High. More to the point, it was the score of that game that Daeshim managed to make confusing.

"What could be easier than telling me the score of a game he just went to?" she asked. "I simply said, 'Daeshim, my man, who won?'"

"Well, what did he tell you, Nicole? Did he tell you who won the game?" Zara asked.

"Well, yes," Nicole responded, "but I wanted to know the score."

"He didn't reveal the score?"

Nicole looked at Mendel as he was the one who asked that question. "He sort of did, but in his own way," she answered.

"And what way was that?" Zara added.

"In a very annoying way!" Nicole huffed. She told the detectives what Daeshim had told her. She knew she was accurate because she'd made him tell her twice. She even wrote it all down, but she still didn't know the score. That's how confusing Daeshim's report had been.

Nicole then summarized what Daeshim told her:

- The Fumblers led after the 1st quarter. The score was 17 to 14.
- In the 2nd quarter, the Fumblers made $\frac{1}{3}$ of the 12 field goals they attempted, 4 of 5 free throws, and 2 of 5 three-point field goals.
- In the 3rd quarter, the Dribblers made 8 of 15 field goals, no three-point field goals, but 8 of 9 free throws. The Fumblers made $\frac{3}{5}$ of the 10 three-point field goals they attempted and one field goal, but missed 2 of 3 free throws.
- In the 4th quarter, the Fumblers scored 2 fewer points than the Dribblers, who made 7 field goals, 4 of 5 free throws, but missed $\frac{2}{3}$ of the 3 three-point field goals they attempted.
- The Dribblers won the game by 4 points.

"There you have it," said Nicole. "That's what he told me."

The detectives agreed that Daeshim's sports report was unnecessarily baffling. But Zara suggested to Nicole that if she sketched a simple scoreboard and filled in the data Daeshim provided, she would be able to figure out the final score herself. Mendel reminded Nicole of the three ways to score in a basketball game.

Basketball Scoring
Field Goal .... 2 pt
Free Throw .... 1 pt
3-Point Field Goal .... 3 pt

Mendel helped Nicole enter what she knew into a scoreboard, like this.

Quarter	1	2	3	4	Final Score
Fumblers					
Dribblers					

**Use Daeshim's details to fill in the scoreboard.
What was the final score?**

Cat Food Case

Finicky Fifi needs more of her favorite feline food.

"My cat Fifi is a fussy eater," Lizzie told Mendel.

That got Digit's attention immediately. Anything about food got his attention—even cat food. (In fact, he didn't think cat food was so horrible. Once he ate from Pebbles's dish and thought the mush was pretty tasty.) That's what he was remembering when Lizzie continued.

"My problem is not that she's a fussy eater, Mendel. I'm a fussy eater. My brother and sister are fussy eaters. My problem isn't about eating, it's about shopping."

Shopping problems are becoming more and more common, thought Mendel. There are so many products to choose from. And there is so much information on the cans, bags, bottles, and boxes. Furthermore, prices can vary a lot for products that are a lot alike.

Lizzie went on to tell Mendel what was troubling her. She explained that her mom had given her the task of buying a can of food for Fifi. The food she was supposed to buy was called Fancy Feline Food. It came in a 15-ounce can. But she was struggling with this job.

"What's causing you so much trouble, Lizzie?" Zara asked as she entered the room.

"The trouble is that my mom gave me very clear instructions. She told me to buy from whichever store had the best price. My mom is like that, always searching for a bargain."

Zara and Mendel understood this because their parents were the same way. "She just wants you to learn to be a smart shopper," Mendel suggested.

"Smart shopping means comparing prices," added Zara.

"I know, I know," replied Lizzie. "And that's where the trouble lies."

"Tell us more, Lizzie," Mendel said.

And she did.

Acme Market:
$1.50 per can—today only

Best Bargains:
3 cans for $4.20!

Cat & Dog Country:
$3.20 per can.
Buy one, get one free!

Daily Dollar: regular price—
$2.00; now $\frac{1}{5}$ off

She explained that she went to the websites of four different stores to check each one's price for Fancy Feline Food. She found that the different stores listed the prices differently, but that all prices applied to 15-ounce cans. She pulled out a notebook in which she'd recorded the prices each store offered. She handed it over to Zara and Mendel to examine.

Digit soon lost interest in the problem since it wasn't really about food. It was about saving money by buying wisely. Since money was never one of his concerns, he trotted over to his water bowl for a refreshing drink. Then he curled up for a nice nap.

But Zara and Mendel were interested. Mendel spoke first. "We see which can is the best buy, Lizzie. And I'll bet you can figure this one out yourself. When you do, you'll make your mom proud and your cat content."

Which store has the best price for the cat food?
How can Lizzie figure it out?

Hiking Head-Scratcher

How can D'Angelo ensure that he reaches a trailhead on time?

Zara and Mendel heard a speeding car come to a screeching halt outside. Something must be up, they thought. They paused the video they were watching to listen. When their doorbell rang, they were sure of it. There stood Alonzo's older brother, D'Angelo, and he was frantic.

"You guys have got to help me! I've got a trail problem."

Before the detectives could even ask what a trail problem was, the frantic fellow continued. "My mom dropped off Alonzo and Zeke at a state park trailhead at noon today. I'm supposed to pick them up when they finish their hike. I know the park, but I don't know what trail they're on. I do know that I'm supposed to show up there at 2 p.m. There are three trails in that park and the trailheads aren't near one another."

Zara and Mendel looked puzzled. "Why don't you just call a brother and ask?" Zara suggested.

"Neither kid has a cell phone," D'Angelo answered.

"Why don't you call your mom and ask her?" Mendel wondered.

"I can't. She's at the dentist getting a cavity filled and won't be able to talk."

The detectives now understood the problem.

"We might be able to help you, D'Angelo," Zara said, "but we'll need some information. To start, tell us about the trails."

D'Angelo described the Rock Creek Trail, the Echo Canyon Trail, and the Lost Lake Trail. He knew that the Lost Lake Trail was a 5-mile loop, that the Echo Canyon Trail was a very steep 4.2-mile loop, and that the Rock Creek Trail was an in-and-out trail that was 2.25 miles in one direction.

Mendel applauded D'Angelo for the good information. But he said that they needed still more before they could help.

"We need to know about how fast the boys walk to figure out which trail they took," Zara said. "Do you know anything about that?"

As it happens, D'Angelo did know something about that.

Digit dropped his ball to listen closely. If he had his way, he'd race along any trail at dizzying dog speed.

"I've walked all these trails myself," D'Angelo replied. He told Zara and Mendel that on flat Rock Creek Trail a young hiker could probably walk at a speed of about 3 miles per hour. The Lost Lake Trail has ups and downs, so he thought the boys could walk it at a speed of maybe 2½ miles per hour. The steep Echo Canyon Trail was more challenging. He estimated that they could keep up a pace of about 1.4 miles per hour.

"You really know these trails, D'Angelo," Mendel remarked in awe.

"Yeah, but how can I know which one will have my worried little brothers at the end of it?"

But no sooner did he say that than D'Angelo's eyes popped open wide. He beamed as if a light bulb had just turned on in his head. Zara and Mendel couldn't help but notice.

"You have all the details you need to collect the boys, D'Angelo. And you've had them all along," Zara said with a grin. "Now, go get your brothers. But this time, drive carefully and stay within the speed limit!"

"Thanks, team," D'Angelo called as he waved and returned to his car. The detectives resumed the video they had been watching. Digit snuggled up to join them.

> **To which trailhead should D'Angelo go to pick up the boys? How do you know?**

Party Animals

It's Digit's birthday, so he's hosting a fabulous party.

Mendel's household was abuzz with excitement. It was Digit's birthday, and he was throwing a party. Actually, he was just attending a party. Because dogs are poor planners and shoppers, Mendel did all the preparation. And since he knew that Digit was truly the doggiest dog in the whole world, he spared no expense. There would be delicious foods, like liver treats and lamb cookies. There would be a puppy play pool and fresh sand to dig in. He'd set up games like Pin-the-Tail-on-the-Kitten and Hide the Bones. Digit was beside himself with joy—his tail twirled like a salad spinner.

Of course, Digit helped with the guest list for he had a pack of friends to choose from. There were other dogs for sure, but he also knew cats, a few ducks, some parrots, a kangaroo, and even a spider—Vinny. Digit really hoped Vinny could make it. He adored that little web master.

Digit ended up with 12 animals (including himself) at the party. Vinny was one of them, which delighted the birthday dog. There were 8-legged, 4-legged, and 2-legged party animals at this shindig. That's quite a few animal feet at one party!

Daeshim came, too, early enough to hide all the bones. But after that job was done, he got bored. The food didn't appeal to him much, so Daeshim began to count animal feet. When he'd finished, he called over to Mendel.

"Say, Mendel, I counted 42 animal feet at this party—not counting humans. Imagine that! Bet you can't figure out how many of each kind without looking."

Come on! Mendel is a math detective, so of course he could. Can you?

> **How many 8-legged, 4-legged, and 2-legged animals were at Digit's party? How could Mendel figure this out?**

Relative Ages

Visiting cousins cause concern for Mischa.

When Mischa came by Mendel's house one morning, he brought two things with him. One was a new chew toy for Digit, and the other was a problem. You might call it an age-old problem. While Digit happily chewed away, Mendel texted Zara to come over.

"What's the good word, Mischa?" she asked as she walked in.

"Cousins," he answered.

"Cousins? That's the good word?" It was Mendel who asked that.

Mischa explained that his cousins from Croatia were coming to visit. He said that this posed quite a crisis.

"What crisis can your Croatian cousins cause?" Zara asked teasingly.

"I'll bet they'll bring a pack of purple pickled peppers," Mendel added.

"Nope. They're bringing me a problem, a problem about their ages."

"You mean you have a problem with their ages?" Zara wondered.

Mischa explained that the problem was that he didn't *know* their ages. The detectives asked why not knowing their ages was a problem.

"It's a problem because my mom wants to get gifts that are right for each one," Mischa answered. "She can't do that if she doesn't know how old they are. We don't see these cousins very often and we know very little about them."

"Why don't you simply ask them, Mischa?" Mendel questioned, while giving Zara a funny look.

It turns out that Mischa's mom did ask, but got a complicated answer.

"Either my Croatian aunt and uncle are fond of funny figuring or they have a weird way of identifying ages in Croatia," Mischa responded.

He told Zara and Mendel what little he knew about his cousins' ages. He said that Ana was the oldest, a teenager whose age is a square number. He explained that Anton is 3 years older than Aleksandar, and added that in 4 years Aleksandar will be 8 years younger than Ana will be then. Mischa said that he didn't know whether that last detail was helpful.

"Told you I had a perplexing puzzle to ponder," Mischa pouted.

"Wow!" said Zara and Mendel in unison.

"Have you left anything out, Mischa?" Zara asked.

Mischa thought for a moment. "Oops! I forgot to tell you that the total of their ages is 35."

While Mischa played tug with Digit and his new chew toy, the detectives put their heads together. That last bit of information allowed Zara and Mendel to confirm their solution to the problem and provide Mischa with the ages of the cousins.

Mischa was grateful for their help. Thanks to the detectives, he could tell his mom what ages to shop for.

"What might she get Ana, the oldest?" asked a curious Zara.

"Oh, probably a basket," he answered with a grin.

"Why a basket, Mischa?"

"It'd be perfect," he responded, "because for spending money, she sells seashells by the seashore."

What are the ages of Mischa's three cousins?

Finders, Keepers

A buried treasure bewilders three friends.

"You don't find buried treasure every day," Charmika announced on the morning bus ride to school. "But we did, the three of us did."

Zara and Mendel were eager to hear more about this, but they had to wait until the end of school. The detectives agreed to do their homework later on so they could meet with Charmika as soon as possible. After all, you don't hear about finding buried treasure every day. Digit was curious, too, since he liked all news about buried stuff.

At 3:30 p.m. sharp Charmika appeared at Zara's house. "I can't wait to hear your treasure tale, but let's wait for Mendel. He should be here any minute," Zara insisted.

Once Mendel arrived, Charmika began her story, which put them on the edges of their seats. Even Digit sat up, his ears upright and his head tilted at the perfect listening angle.

Charmika first set the scene. She, Leo, and Paco were hiking in the nearby national forest. Paco's mom had dropped them off there on Sunday afternoon after lunch. The plan was to call Paco's mom to pick them up after they hiked the beautiful lakeside trail.

They had walked for about half a mile when Leo spotted a lesser-used trail leading off to the right. He urged the others to take it, and they cautiously agreed. It led them across a meadow and then curled into a dense, dark wooded area. They had been walking that overgrown trail for about 20 minutes when Leo noticed a small mound of earth by a huge oak tree.

"Look at that!" he said.

"It's a pile of dirt, just leave it. What's so special about dirt?" Charmika asked.

"It was made by someone. Maybe there's something buried beneath it," Leo answered as he walked toward it. Charmika and Paco rolled their eyes, but followed. The two watched Leo use his hands and a stick to level the mound and dig underneath. Within moments, he felt something that wasn't dirt. He gasped.

"There's something buried here," he yelled, and began to dig with excitement. Soon he clutched something in his hand and revealed it to his friends. It was a bag, an ordinary but very dirty cloth bag. And it was heavy.

The bag was heavy because it was filled with coins— all quarters. Lots of them. Leo placed the bag of money into his backpack, and the three completed their hike. Paco called his mom, who arrived within moments. As they waited, the friends agreed to keep the money— they didn't know what else to do.

Charmika reported, "We all agreed that since it was Leo's idea to take that other path, and since he was the one who found the mound and unearthed the bag, it should be his money to keep or divide up. He chose to share it. He decided to give $\frac{1}{3}$ of the money to Paco. He will give $\frac{1}{4}$ of it to me. And he'll keep the rest for himself. Paco gets a little more because he wasn't the one who'd said to ignore the pile of dirt."

Paco's mom dropped Charmika and Leo off at their houses. When Leo got home, he ran to his room, dumped the quarters onto the bed and counted them. It took him some time to divvy up the loot according to the plan. When he finished, Leo texted his pals.

My take is $25. I'll give you your shares tomorrow.

"Great story, Charmika!" Mendel exclaimed. "But how can we math detectives help?"

Charmika sheepishly said that when she told her parents about the treasure, they wanted to know how much money was in the bag and how much she would get. "I didn't know what to say," she admitted. "I didn't want to ask Leo, but I do want to know the answer. Can you guys help me?"

They could and did. Mendel used some guesswork and his keen number sense until he figured out the solution. Zara used algebraic thinking.

> **How much money was in the bag?**
> **How much did Charmika and Paco each get?**
> **How did the detectives figure it all out?**

Tech Time-Crunch

Friends help to finish a project on time.

Every spring, Zara and Mendel's school holds a technology fair. Students work really hard to design and build all sorts of cool gizmos. You might see projects with parts that bubble, hum, clatter, or beep. You might be amazed by robots that can feed a cat or fold clean laundry. Teams compete to create the most unique or ingenious contraptions. Prizes are awarded to the best of the best.

The winning project from last year's tech fair was a remote-controlled dolphin that spit fresh water from its mouth onto thirsty houseplants and into empty pet bowls. Daeshim guffawed for at least 10 solid minutes after hearing about that one.

So when Ingrid and Felix appeared at Mendel's house with a tech-fair problem, he knew he'd have to think hard with such serious customers on his hands.

"We've got something special up our sleeve, Mendel," Ingrid began.

"It's a sure-fire winner," added Felix. "But we have a problem— a worker/time problem."

Mendel wondered what they meant by "a worker/ time problem." So did Zara, who had walked in during that part of the conversation. Digit was also there. But he never did any work so had never had a work problem of any kind. He found licking his paws far more interesting.

"What do you mean, guys?" Mendel asked.

Ingrid explained that she and Felix were rushing to complete their awesome invention on time. But time was running out. "We need more workers, that's all there is to it," Felix admitted.

Ingrid filled in the details. "Here's our problem exactly," she said. "We've already worked for two weeks, and estimate that it will take 30 more hours to finish. But with so much else to do, I can put in only 11 more hours and Felix can work only 10 more hours. So unless time stops—which it probably won't—we're in trouble."

Zara and Mendel agreed that the two young techno-wizards had a time problem on their hands. Felix added, "Since we've fallen behind we want to ask some of our pals to pitch in. We know we can quickly teach them to do what we need, whether it's measuring, building, painting, or testing."

"We estimate that it's reasonable to ask each friend for 1½ hours of their time. They'll do it, we think. It might cost us cupcakes all around, but we think it's our only hope," Ingrid explained.

"But how many kids will we need?" Felix asked. "We're stumped."

"Yeah, how many cupcakes will Felix need to bake?" added Ingrid slyly.

Both detectives were taking notes as the scientists provided key information. They compared calculations while Digit, having finished with his paw, now scratched his left ear. Within moments Zara and Mendel had an answer for Felix and Ingrid.

How many friends will Ingrid and Felix need to help them complete their project on time?

Fish Scales

How can a fish tale become even fishier?

Liane has an uncle named Sid, who likes to go fishing. He also likes to tell tall tales about it. Liane thinks that this is all well and good until Uncle Sid gets annoying.

"I love my uncle but he can be so aggravating," she told Zara one Sunday morning. "I don't mind a good whopper. But his whoppers can be impossible to understand. You can't even tell how much he is exaggerating!"

Zara laughed out loud. She guessed that Liane had just heard one of Uncle Sid's annoying explanations and that she was about to hear all about it. And she was right.

"Yesterday Uncle Sid went fishing on Lake Luzerne with his friend Dave," she told Zara and Mendel, who had just joined them in the room.

"Hi, Liane, what did I miss?"

Zara brought Mendel up to speed, then invited Liane to continue her story.

"I asked Uncle Sid a simple question—how much did the biggest fish he caught that day weigh? I naturally expected his usual stretching of the truth. But his answer made my head spin like the reel on his fishing rod."

Zara and Mendel chuckled at that image. "Go on," Mendel encouraged.

"So Uncle Sid told me that the biggest fish Dave caught weighed 4½ pounds. He rolled his eyes and cupped his hands to show that he thought that fish was guppy-sized. Then he snickered, adding that Dave's biggest catch weighed ½ pound less than ⅓ the weight of his own biggest catch that day. Do you understand what I mean by 'annoying'? I still don't know the weight of his staggeringly enormous fish!"

Mendel was recording the data Liane had given and was making calculations. "Well, I now can tell you the weight of that fish, Liane," he stated with certainty.

> **How much did Uncle Sid's biggest catch of the day weigh?**

Meeting Marvin

The detectives help Ruby pick up a visitor from another time zone.

Ruby was super excited when she called Zara. She had been excited all day because Marvin was coming to visit. Marvin was her favorite cousin. But he lived so far away—nearly across the country—so visits were rare. But Marvin was on his way. He had texted her from his granddad's car on their way to the airport in Denver, Colorado.

Ruby told Zara that Marvin's granddad picked him up in plenty of time to make the 8:00 a.m. flight to New York, where she lived. But alas, the flight was delayed. She said that Marvin had texted again to let her know that his plane would arrive behind schedule. Then he texted once more just before his plane took off—only 45 minutes late. He added that the pilot promised to shave 15 minutes off the originally scheduled flight time of 4 hours, 40 minutes. Ruby was grateful that the pilot would try to make up some of the lost time.

Zara listened carefully to Ruby's story. She learned that Marvin had also called Ruby the night before. He'd told her that once he landed, it would take him another 45 minutes to collect his luggage and walk to the bus stop outside the terminal.

"I know that bus," Ruby told Zara. "Its route into the city is 18 miles long and costs $26. Oh, and it takes 30 minutes, unless there's a traffic jam."

Zara was waiting to hear whether there was a problem for her to solve. She figured it would likely have to do with travel time. She didn't have to wait long.

"So here's my problem, Zara," Ruby said. "I need to tell my mom when we should leave our house to pick up Marvin at the bus station."

With that, Zara was silent for a moment. Then she spoke.

"I need four things before I can help you, Ruby. First I need to know how long it takes your mom to drive from your house to the bus station."

"Not very long—10 minutes door to door," answered Ruby. "What else?"

Zara also needed a pencil, some paper, and a time-zone map of the United States.

Zara did some figuring while Ruby watched. In just a few minutes Zara had figured out when Ruby and her mom should leave to meet cousin Marvin at the bus station.

What time did Zara tell Ruby to leave home?
How did she figure it out?

The Broom King

The solution to this mystery can be as sweeping as your imagination.

You remember Alex—the girl who works at her Uncle Max's awful ice-cream shop? It turns out that she has more than one uncle who owns a store. And when Uncle Arnold heard about the fabulous sign she made for Max, he wanted to hire her as well. When Alex heard that, she went straight over to Zara and Mendel.

"What's the uncle problem this time, Alex? Did Max come up with a new, even more disgusting ice-cream flavor?" Mendel inquired.

"Yeah, like liver and pickles?" chuckled Zara.

"Different uncle, different problem," Alex answered, and began to explain. She said that Uncle Arnold has a store that sells brooms, only brooms. She said that the store—unexpectedly called BROOM CLOSET—sells more brooms than any other store in the area.

"So let me guess—he wants you to make a broom sign? It better be tall and thin or wide and narrow," noted Mendel with a wry grin.

"Much worse. He wants me to come up with a clever marketing strategy. He wants customers to know how many brooms he sells, but he wants me to do this in a creative and unexpected way."

"Do you have any ideas yet, Alex?" Zara asked.

She did have an idea, an idea she really liked. The problem was figuring out how to put her idea into practice.

"Tell us your idea and maybe we can help," Mendel offered.

Her idea was to express how many brooms Uncle Arnold's store sells in a year in a catchy way.

"You mean without simply naming a number—like 10,000?" Zara wondered. "Or would that just be too boring?"

"Yes, it would be," Alex answered. "Uncle Arnold expects more of me. So I mopped my brow and came up with what I hope is a tempting idea. What if I could describe the number of products sold in terms of how long a line they would make if all of them were placed end-to-end? I could use sales figures from the most recent week or month—they have been terrific!—to make Uncle Arnold's case that he is undoubtedly the Broom King."

"Wow! You mean, like 'they would make a line five miles long'?" Mendel asked.

"Sort of, but better," answered Alex. "What if I could say that they made a line 'from Boston to Miami'? Or 'across the Atlantic Ocean'? Or 'up one side of Pike's Peak and down the other'? Wouldn't that be wildly creative?"

Zara and Mendel agreed that it would. "It's an awesome idea, Alex, but how will you figure out what to say?" Zara asked.

"That's why I'm here, pals. Can you help me get started?"

They could and did. The three worked together to come up with a strategy to estimate and describe broom sales that would sweep away the competition.

> **Describe the steps you could take
> to help solve Alex's problem.**

Certificate of Recognition
for Excellence as a Math Sleuth

is hereby presented to

_____

Congratulations!
—Zara, Mendel, Daeshim & Digit

Certificate of Recognition
for Excellence as a Math Sleuth

is hereby presented to

_____

Congratulations!
—Zara, Mendel, Daeshim & Digit